PLANT GROWTH

Edited by **Everlon Cid Rigobelo**

Plant Growth
http://dx.doi.org/10.5772/62601
Edited by Everlon Cid Rigobelo

Contributors
Mehi Lal, Saurabh Yadav, Vivek Singh, M Nagesh, Munazza Gull, Arturo Herrera, Fauzia Chattha, Munawer Ali Munawer, Samina Kousar, Mehr Un-Nisa, Xingxue Huang, Aihua Wang, Bincai Wang, Guolin Zhou, Chufa Lin, Yaohua Deng, Ruahua Zhang, M. Angeles Forner-Giner, Mary-Rus Martinez-Cuenca, Amparo Primo-Capella, Jie He, Waleed Abdulkhair, Metin Turan, Nurgül Kıtır, Emrah Nikerel, Ertan Yıldırım, Şefik Tufenkçi, Adem Günes, Summy Yadav, Kamal Dutt Sharma, Michael Lieber, Meixue Zhou, Ibrahim Danasabe Ahmed, Matthew Harrison, Holger Meinke

CBS, Edition **2017**

Published by InTech
Janeza Trdine 9, 51000 Rijeka, Croatia

Publishing Process Manager Dajana Pemac

Technical Editor SPi Global

Cover Designer InTech Design Team

Additional hard copies can be obtained from orders@intechopen.com

Plant Growth, Edited by Everlon Cid Rigobelo
p. cm.
Print ISBN 978-953-51-2771-0
Online ISBN 978-953-51-2772-7

Contents

Preface VII

Section 1 Biology 1

Chapter 1 **The Use of Bio-Agents for Management of Potato Diseases 3**
Mehi Lal, Saurabh Yadav, Vivek Singh and M Nagesh

Chapter 2 **Plant Health 19**
Munazza Gull

Chapter 3 **The Organic Amendment Improve the Yield and Quality of Vegetable 35**
Xingxue Huang, Aihua Wang, Bincai Wang, Guolin Zhou, Chufa Lin, Yaohua Deng and Runhua Zhang

Chapter 4 **Plant Pathogens 49**
Waleed M. Abdulkhair and Mousa A. Alghuthaymi

Chapter 5 **Making Soil More Accessible to Plants: The Case of Plant Growth Promoting Rhizobacteria 61**
Metin Turan, Nurgül Kıtır, Ülker Alkaya, Adem Günes, Şefik Tüfenkçi, Ertan Yıldırım and Emrah Nikerel

Section 2 Physiology 71

Chapter 6 **The Unsuspected Intrinsic Property of Melanin to Transform Light into Chemical Energy and the Seed Growth 73**
Arturo Solís Herrera

Chapter 7 **Coumarin-Based Heteroaromatics as Plant Growth Regulators 91**
Fauzia Anjum Chattha, Mehr-un-Nisa, Munawer Ali Munawer and Samina Kousar

Chapter 8 **Influence of Rootstock on Citrus Tree Growth: Effects on Photosynthesis and Carbohydrate Distribution, Plant Size, Yield, Fruit Quality, and Dwarfing Genotypes 107**
Mary-Rus Martínez-Cuenca, Amparo Primo-Capella and Maria Angeles Forner Giner

Chapter 9 **Root Growth, Morphological and Physiological Characteristics of Subtropical and Temperate Vegetable Crops Grown in the Tropics Under Different Root-Zone Temperature 131**
Jie He

Chapter 10 **Molecular and Morphophysiological Analysis of Drought Stress in Plants 149**
Summy Yadav and Kamal Dutt Sharma

Chapter 11 **Barley Phenology: Physiological and Molecular Mechanisms for Heading Date and Modelling of Genotype-Environment-Management Interactions 175**
Ahmed Ibrahim, Matthew Harrison, Holger Meinke and Meixue Zhou

Section 3 Genetic 203

Chapter 12 **Toward a New Paradigm for the Evolution of Developmental and Growth-Pattern Systems in Plants and Animals 205**
Michael M. Lieber

Preface

Since the appearance of human beings, we are dependent on plant growth. Many studies have helped us to know a little about this phenomenon and yet there is a lot to learn. The present book is a great opportunity to deepen our knowledge in this subject. This book brings twelve chapters distributed into three sections: Biology, Physiology, and Genetics. In the Biology section, the chapters have addressed the use of bio-agents for management of potato diseases, plant health, organic amendment, and plant pathogens. In the Physiology section, the chapters have addressed the property of melanin to transform light into chemical energy, coumarin-based heterocromaties as plant growth regulators, the influence of rootstock on citrus tree growth and root growth, and molecular and morphophysiological analysis of drought stresses in plants and barley phenology. Finally, the Genetics section has a chapter which has addressed a model plant-like organisms.

This book has been designed for researchers, students, and people who never tire of learning.

I would like to thank Dr. George Lazarovits, a great researcher who is working with plant growth-promoting bacteria for 40 years. I would like to thank my wife Fernanda, my daughter Maria Eduarda, and my son João Henrique for making my life happier.

Everlon Cid Rigobelo
Plant Production Department at UNESP University
Jaboticabal, Sao Paulo
Brazil

Section 1

Biology

Chapter 1

The Use of Bio-Agents for Management of Potato Diseases

Mehi Lal, Saurabh Yadav, Vivek Singh and M Nagesh

Additional information is available at the end of the chapter

http://dx.doi.org/10.5772/64853

Abstract

Potato is an important food crop in the world as well as in India. It is being affected by different pathogens, viz. fungi, bacteria, viruses and nematodes. These pathogens may cause significant yield losses of the crop, if proper protection measures have not been applied. Among potato pathogens, *Phytophthora infestans, Alternaria solani, Rhizoctonia solani* and *Fusarium* spp. are the major pathogens in the fungal group, whereas *Ralstonia solanacearum, Pectobacterium* spp. and *Streptomyces* spp. are in the bacterial group. For management of these pathogens, various methods, that is, chemical control, biological control, resistant varieties, cultural control and physical control, are applied. Resistant varieties are the best and cheapest method for managing the diseases. However resistant varieties are break down their resistant over the years and moreover against some pathogen absolute resistant are not available. Chemical management is the second best option for managing the diseases, due to continuous and irrational use of the chemicals; pathogens have developed resistance against certain class of fungicides/bactericides. Moreover, these chemicals also assist in environmental pollution and toxicity in the produce. Bio-agents are naturally occurring living organisms, which are found in rhizosphere, phylloplane, etc. These bio-agents help in not only managing the diseases but also increasing the crop yield. Therefore, the use of bio-agents for biological management of potato crops is the focused research area worldwide.

Keywords: Bioagents, potato, diseases, management, bacteria, fungi

1. Introduction

Potato originated in the hills of Andes and Bolivia in South America. It was introduced into Europe by Spaniards in the second half of the sixteenth century, from there it spreads throughout Europe and rest of the world in the mid-seventeenth to mid-eighteenth century. In India, it was introduced by Portuguese in the seventeenth century. Potato is the most important crop in the world. It is affected by various diseases and pests. Diseases are the major cause of concern for reducing the economic yield and affecting status of the potato growers. Major diseases of potato are late blight, early blight, black scurf, dry rot, etc. in the fungal group, whereas bacterial wilt, soft rot/blackleg of potato and common scab in the bacterial group. Sometime these diseases may cause losses up to 75%. Potato diseases can be managed by various methods, viz. chemical control, cultural control, biological control, physical and resistant varieties. Generally, chemical control is used for managing the diseases at large scale. Due to use of chemicals (fungicides/antibiotics) for longer periods for managing the disease, it was observed that pathogens have developed resistance against certain chemicals, besides also enhanced the toxicity in the environment. To avoid development of resistance in pathogens and toxicity in the environment, the use of bio-agents/biological control is the best option. In a simple way, biological control can be defined as the partial or total inhibition or destruction of pathogen population by other microorganisms. Broader way, Baker and Cook (1974) defined this as the reduction of inoculum density or disease-producing activities of a pathogen or parasite in its active or dormant state, by one or more organisms, accomplished naturally or through manipulation of environment, host or antagonist or by mass introduction of one or more antagonist [1]. The first experiment in biological control with antagonists was conducted by GB Sandford in Canada [2]. Different mechanisms of biological control of pathogenic fungi have been suggested, including microbial competition, antibiosis, hyperparasitism and induction of systemic acquired resistance in the host plants [3]. Bio-agents have remarkable capacity of multiplication; thus, when applied they multiply in exponential ratio and even can overcome stress conditions by forming thick-walled spores [4]. Recent years have witnessed the increasing popularity of biological control agents as an alternative to fungicides [5]. *Trichoderma* species as biocontrol agents (BCAs) was recognized for the first time by Weindling [6]. *Bacillus* spp. and *Pseudomonas* spp. have been used in biological management of the potato diseases. Bio-agents are effective against seed and soil-borne plant pathogens. The biological control of soil-borne plant pathogens has drawn much attention in the past few decades and is currently considered as a promising alternative to synthetic pesticides because of its safety for the environment and the human health [7]. Plant growth promoting rhizobacteria (PGPR) and vesicular-arbuscular mycorrhizae (VAM) are known to minimize plant diseases and increase crop yield. Biocontrol applications on potato plants require a better knowledge of its beneficial fungal partners. This kind of microbial community has been poorly studied, particularly because in vitro cultivation of mycorrhizae remains difficult [8]. Biocontrol agents are an important component especially in the organic cultivation of potato. Biological control of major fungal and bacterial diseases of potato is discussed in the following sections:

2. Late blight of potato

The late blight disease caused by oomycete has a great importance in the history of plant pathology. Initially, its causal organism was reported *Botrytis infestans* in 1845 by C. Montagne, a retired French army doctor who had devoted much of his life to the study of fungi. About 30 years later, German scientist Anton de Bary renamed it as *Phytophthora infestans* (Mont.) de Bary [9]. During 1844–1845, the entire crop across Europe, especially in Ireland, was destroyed prematurely leading to worst ever famine the 'Irish Potato Famine' [10]. One million people died of starvation and another million migrated to USA and other parts of the world. The late blight fungus co-evolved with potato in Central and South America and subsequently spread to other parts of the world mainly through infected seed tubers. Late blight was recorded in India for the first time between 1870 and 1880 in the Nilgiri Hills [11]. Under subtropical plains, it was first observed in 1898–1900 in Hooghly district of West Bengal [12]. In the northern part, it appeared for the first time in 1883 in Darjeeling and spread rapidly to adjoining hills [13]. *Phytophthora infestans* caused late blight diseases in potato and tomato crops worldwide. It not only caused economic losses of yield but also the quality and quantity of the crop. Recently, reduction in 10–15% yield was expected at national level (India) due to occurrence of late blight disease [14]. *Phytophthora infestans* is highly researchable pathogen in plant diseases. The worldwide late blight disease is re-emerging; therefore, this disease is constantly observed by the late blight researchers [15]. Late blight affects all plant parts, especially leaves, stem and tubers. Whitish mycelium appears on lower leaves under humid conditions and is the most important symptom. Light brown lesions develop on stem and petioles, and rusty brown discolouration of the flesh is the typical symptom of late blights on potato tubers. The pathogen is mainly seed borne in nature but also soil borne in some cases. Management of late blight through eco-friendly means of applying botanicals has been initiated in European and American countries during the past years of the twentieth century [16, 17]. Of 100 species in 54 plant families tested, leaf extracts from onions, garlic, *Malus toringo, Reynoutria japonica* and *Rheum coreanum* inhibited mycelial growth of *P. infestans. M. toringo* extracts strongly inhibited *P. infestans* and was effective in controlling late blight also [18]. Some antifungal compounds reported from botanicals against late blight of potato [19]. The antagonist *Bacillus subtilis* B5 was found effective in inhibiting the growth of *P. infestans* [20]. Integrated management of late blight, using two sprays of *Bacillus subtilis+ Trichoderma viride* and one spray of fungicides, at the onset of disease is found to be effective for managing late blight of potato [21]. Rhamnolipid is a class of glycolipids, which is produced by bacteria. Rhamnolipid-based formulation (0.25%) from *Pseudomonas* spp. was tested under field trials at three different locations. The terminal disease severity in rhamnolipid formulation was 45% (compared to 100% in control), 47.5% (against 92.5%) and 59.2% (as against 76.64%) at Modipuram, Lavad, (Meerut) and Jalandhar, respectively [22]. Certain microorganisms in the phyllosphere were antagonistic to *P. infestans*, which included the yeasts *Sporobolomyces* spp., *Acetobacter* spp., isolates of *Pseudomonas* spp. and *Bacillus* spp. [23, 24]. *Bacillus* sp. inhibited mycelial growth of seven plant pathogenic fungi *in vitro* and *in vivo*, and the same bacterium protected tomato plants against *P. infestans* [25]. Various bio-agents, including a bacterium (*Serratia* sp.) and four fungi (*Trichoderma* sp., *Fusarium* sp. and two *Penicillium* spp.), were evaluated against *P. infestans* on tomatoes under field conditions at Costa Rica, and it was reported that *Penicillium* reduced the lesion area/plant between 8 and 40% [26]. One hundred and twenty-two microorganisms isolated from the phyllosphere of potatoes and only 23 were effective microorganisms (spore-forming and non-spore-forming bacteria, yeasts and fungi) in dual cultures with different patterns of inhibition of *P. infestans* [27]. Various naturally occurring microorganisms, that is, *T. viride, Penicillium viridicatum, Penicillium aurantiogriseum, Chaetomium brasiliense* [28], *Acremonium strictum* [29], *Myrothecium verrucaria* and *P. aurantiogriseum* [30], showed antagonistic effect against *P. infestans*. The antagonistic activities of *Pseudomonas fluorescens, Pseudomonas* sp. *Aspergillus flavus, Aspergillus niger, Penicillium* sp., *Trichoderma virens* and *Trichoderma harzianum* showed positive inhibition of mycelial growth of *P. infestans, Fusarium* spp. and *Rhizoctonia solani* under *in vitro* conditions **Table 1** [31]. One hundred fifty-two endophytic fungi were isolated from healthy tissues of vegetable plants, and only 23 (15%) isolates showed anti-oomycete activity against tomato late blight and in *vivo* [32].

Treatments	*Rhizoctonia solani*		*Fusarium* sp.		*Phytophthora infestans*	
	Inhibition of growth (%) after 3 days over control	Bell's rating	Inhibition of growth (%) after 8 days over control	Bell's rating	Inhibition of growth (%) after 12 days over control	Bell's rating
Pseudomonas sp.	32.22	4	47.16	3	55.68	2
P. fluorescens	39.25	4	53.01	2	53.40	2
A. flavus	39.44	3	43.77	3	59.00	2
A. niger	56.48	3	50.18	2	61.36	3
Penicillium sp.	37.22	3	63.01	2	59.00	2
T. virens	42.77	2	52.64	2	64.77	2
T. harzianum	46.11	2	57.16	2	63.63	2
CD (0.05)	8.80		5.97		2.59	

Table 1. Antagonism between bio-agents and potato pathogens.

Naturally occurring surface active compounds derived from microorganisms are called biosurfactants. These are amphiphilic biological compounds produced extracellularly as part of the cell membrane by a variety of bacteria, yeast and fungi [33]. Research on biosurfactants used as a biocontrol, particularly in potato against *P. infestans*, has initially started in India under PhytoFura network project. Biosurfactants can be used as alternatives to chemical surfactants as their capability of reducing surface and interfacial tension with low toxicity, high specificity and biodegradability make them important for inhibiting pathogens. The metabolite of biosurfactant-producing microorganism (*Pseudomonas aeruginosa*) has shown high efficacy against *P. infestans* under *in vitro* conditions [34]. Ninety-five isolates of bacteria were tested for their biosurfactant as well as biocontrol activity against *P. infestans*. Results revealed that only 15.8% isolates showed biosurfactant activity and only five isolates were found to be effective against *P. infestans* for biocontrol properties. Amongst highest effective was *P. aeruginosa*, which was tested in different forms, viz. bacterial cells, culture filtrate and formulation against *P. infestans* on whole plant method and lowest disease severity (9.44%) recorded with culture filtrate excluding mancozeb treatment mentioned in **Figure 1** [35]. Biosurfactants produced by bacteria, yeasts and fungi can serve as green surfactants. However, large-scale production of these molecules has not been realized because of low yields in production processes and high recovery and purification costs [36]. The best antagonistic activity against *P. infestans* is observed in the genera of *Pseudomonas* and *Bacillus* as they produce a wide range of antibiotics and biosurfactants and can be used as alternatives to chemical surfactants [37].

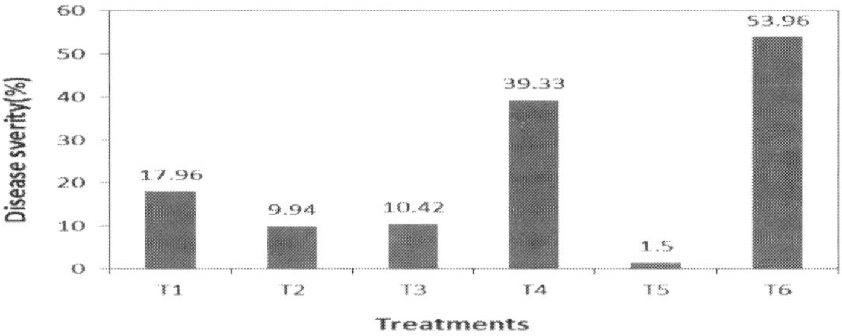

Figure 1. Effect of different forms of bio-agents on late blight development using whole plant method T1—bacterial suspension of *Pseudomonas aeruginosa* 1, T2—culture filtrate of *P. aeruginosa* 1, T3—bioformulation of *P. aeruginosa* 1, T4—Talc powder, T5—Mancozeb (0.2%) and T6—distilled water spray (control).

3. Early blight of potato

Early blight of potato caused by *Alternaria solani/A. alternata.* The symptom of this disease is dark brown to black lesions with concentric rings, which produce a 'target spot' effect. Symptoms are initially observed on older leaves and weaker plants. *A. solani* is a polycyclic pathogen as many cycles of infection are possible during a season [38]. The antimicrobial activity of six plant extracts from *Ocimum basilicum* (Sweat Basil), *Azadirachta indica* (Neem), *Eucalyptus chamadulonsis* (Eucalyptus), *Datura stramonium* (Jimsonweed), *Nerium oleander* (Oleander) and *Allium sativum* (Garlic) was tested for managing *Alternaria solani in vitro* and *in vivo.* The results revealed that the highest reduction of disease severity was achieved by *A. sativum* at 5% concentration and the lowest reduction was obtained when tomato plants were treated with *O. basilicum* at 1% and 5% concentration [39]. *T. viride* (0.5%) was found to be effective against early blight of potato for reducing disease intensity under field conditions [40]. The bio-agents *T. harzianum* and *P. fluorescens* (seed treatment + foliar spray) were effective in reducing the disease intensity of early blight of potato and also increasing tuber yield [41].

4. Black scurf of potato

Black scurf is an important disease of potato in the category of soil- and tuber-borne diseases. Infected seeds are the main sources of infection [42]. It affects roots, stolen, stems and tubers. The disease has two phases, viz. stem canker and black scurf. Stem canker phase is the girdling on the stem with brown colour and sometime upward rolling of the leaves also observed. Black scurf phase is formation of sclerotia on the surface of the tubers. This phase is more common in the field, particularly at the stage of plant senescent. *Rhizoctonia solani* has wide host range, and it is soil and seed borne in nature. Seed treatment by chemicals is effective against seed borne. However, biological control is a better option than chemical control in relation to creating pollution in the environment. The seed treatment with 1.5% boric acid followed by an application of a *T. viride* formulation containing 1×10^7 c.f.u./g @ 4.5%/kg seed tubers at planting reduced the disease to level achieved with 3% boric acid spray [43]. Out of 28 isolates, nine bacterial strains were found to be antagonistic *in vitro,* reduced the fungal growth and caused the lysis of sclerotia of *R. solani* in a dual culture assay as well as in an extracellular metabolite efficacy test. The selected antagonistic bacteria were also characterized for growth promoting attributes, that is, phosphate solubilization, nitrogen fixation and indole acetic acid production. Biocontrol efficacy and per cent yield increase by these antagonists were estimated in a greenhouse experiment, and results showed that two *Pseudomonas* spp. StT2 and StS3 were the most effective with 65.1 and 73.9% biocontrol efficacy, as well as 87.3 and 98.3% yield increase, respectively [44]. Potato seed treatment showed higher efficacy than the soil drenching when both ways (seed treatment and soil drenching) separately used with fungal and bacterial bio-agents to manage the black scurf of potato [45]. The interaction of PGPR (*Bacillus* spp.) with potato seeds or vegetative parts showed promising antagonism through producing siderophore and antibiotics against black scurf and stem canker diseases of potato caused by *R. solani,* thereby resulting in increase of potato yield. The effectiveness of PGPR strain (*Bacillus* spp.) was observed in improving the yield of potato in greenhouse and in the field conditions [46]. Seed treatment by *T. viride* showed less disease index of black scurf of potato against control [47]. Whereas, when *T. viride* including other bio-agents compared, it was found that *T. harzianum* significantly inhibiting the mycelia growth of *R. solani* [48]. Bio-agents not only reduce the disease incidence but also increase the crop yield, compared to without the use of bio-agents [49]. Sunhemp and maize green manuring reduced the disease incidence of black scurf of potato [50]. Chopped leaf matter of brassica crops and barley inhibited growth of Rhizoctonia, while Indian mustard almost completely inhibited the mycelial growth of *R. solani* [51]. The antagonistic effect of microorganisms was evaluated after adding rhizospheric extracts of maize, oat, barley and grass on Rhizoctonia. It was observed that extracts from maize and grass rhizosphere were most antagonistic [52]. The antifungal efficacy of six botanical extracts and two bio-agents, viz. *T. harizanum* and *T. viride,* were evaluated *in vitro* against sclerotial isolates of *R. solani* causing black scurf of potato through food poison and dual culture technique, increasing concentration from 5 to 15% of botanical extract suppressed the mycelial growth of all isolates. Among the tested bio-agents, mycelial growth inhibition of *R. solani* isolates was recorded in the case of *T. harzianum* (up to 72.72%) and *T. viride* (up to 56.80%) [53].

5. Fusarium wilt/dry rot of potato

Fusarium dry rot is an important post-harvest disease of potato tubers. This disease is distributed worldwide and occurs wherever potatoes are grown [54]. *Fusarium* spp. cause fusarium wilt in the field and under storage it causes dry rot of potato. *T. harzianum* (ANR-1) isolate was found to be effective in inhibiting the radial mycelial growth of *Fusarium oxysporum* f. sp. *lycopersici* (53%). Under greenhouse conditions, the application of *T. harzianum* (ANR-1) exhibited the least disease incidence (15.33%) and also found stimulatory effect on plant height (73.62 cm) and increased the dry weight (288.38 g) of tomato plants in comparison to other isolates and untreated control [55]. Immature crop plant amendments, viz. pearl millet, sesbania, sunhemp, maize and eucalyptus leaves, are used against fusarium wilt of potato. Among them, eucalyptus leaves and maize showed maximum suppressive and least was sesbania [56]. The combined effect of antagonists (*Trichoderma* and *Pseudomonas*) with modified montmorillonite particles (Mod- MMT) against *F. oxysporum* f. sp. *tuberose* causes wilt of potato, showed less disease incidence and also enhanced plant height, fresh and dry weight, number of tubers/plant and weight of tubers [57]. Application of *Trichoderma koningii* and *Bacillus megaterium*, alone or in combination, seven days earlier than soil infestation with *F. oxysporum* and/or the mixed population of *Meloidogyne* spp., significantly reduced *Fusarium* wilt disease incidence and nematode infection on potato and improved plant growth components under greenhouse conditions. Generally, the mixture of the two biocontrol agents was more effective in controlling the plant disease and improving plant growth components than either of the two organisms used singly [58].

6. Common scab of potato

Potato common scab caused by pathogenic *Streptomyces* spp. is a serious disease in potato production worldwide. It occurs throughout the potato-cultivating regions of the world and is most prevalent in neutral or slightly alkaline soils, especially during dry years [59]. The disease symptoms are small brownish, shallow, raised or sunken and mostly appeared on tubers. The pathogen is both seed and soil borne. The pathogen is survived for longer period in the infected plant debris and soil. Biological control of common scab is one of the attractive approaches which can develop naturally in potato fields owing to antagonistic microorganisms and reduce the severity of disease [60, 61]. Three antagonistic fungi, that is, *T. harzianum, Penicillium digitatum* and *Aspergillus flavus*, were evaluated for biological management of common scab of potato. Results revealed that lowest disease incidence was observed with *T. harzianum* [62]. *Pseudomonas mosselii* when applied with vermicompost gave the best plant growth and yield along with maximum reduction in scab incidence and scab index [63]. Most actinomycete isolates derived from the Rice bran-amended soil showed antagonistic activity against pathogenic *Streptomyces scabiei* and *Streptomyces turgidiscabies* on R2A medium. Some of the *Streptomyces* isolates showed positive results when they were inoculated onto potato plants in a field condition. These results suggest that Rice bran amendment increases the levels of antagonistic bacteria against pathogenic strain of Streptomyces in the potato rhizosphere

[64]. Phage therapy is a new method to manage plant pathogens. Phage therapy has allowed disinfection of *S. scabiei*-infected seed potatoes and reduced tobacco bacterial wilt due to *R. solanacearum* by co-application with an avirulent strain of this bacterium [65, 66]. The culture broth of *Bacillus* sp. *sunhua* had a suppressive effect on common scab disease in a pot assay, decreasing the infection rate from 75 to 35% [67]. Non-virulent potato isolates of *Streptomyces* spp., with antagonistic activity higher than PonSSII, significantly reduced scab in pot experiments. Two non-pathogenic strains of *Streptomyces*, viz. *S. diastatochromogenes* strain PonII and *S. scabies* strain Pon R found to be effective against the pathogenic strain of *S. scabies* of potato in 4-year field experiments [68, 69].

7. Black leg of potato

Black leg of potato caused by different species of bacteria, viz. *Pectobacterium* spp. (*Erwinia* spp.) and *Dickenya* spp. [70]. Both are pectinolytic in nature and represent a significant threat for seed potato production in Europe. *Dickeya* spp. induce various symptoms such as plant wilting, stem rot (blackleg) and tuber soft rot [71]. The bacteria live over in soil in decaying plant debris and sometimes in seed tubers. *Pseudomonas* spp. and *Bacillus* spp. were evaluated against *Pectobacterium* spp. The antagonistic properties of different *Pseudomonas* spp. strains, such as iron competition, 2,4-diacetylphloroglucinol (DAPG) antibiotic synthesis via pyoverdine and pseudobactin production and their related receptors, were found to be the means of protection [72, 73] against *Pectobacterium* spp. *Bacillus subtilis* strains were tested for the control of potato diseases caused by *Pectobacterium* spp., and results revealed reduced maceration symptoms in planta [74]. A bacteriocin-like substance produced by *Bacillus licheniformis* P40 was bactericidal to *Pectobacterium carotovorum* subsp. *carotovorum*. This substance interacted with cell membrane lipids, provoking lysis of *P. carotovorum* subsp. *carotovorum* cells. It was also effective in protecting potato tubers against soft rot under standard storage conditions [75]. Different strains of *P. fluorescens* were used to protect wounds and cracks on tubers from colonization by *Pectobacterium atrosepticum*. Application of individual and combinations of strains reduced the contamination of potato tuber peel by 85% and 60–70%, respectively, indicating the potential of *Pseudomonas* spp. for controlling soft rot caused by *Pectobacterium atrosepticum* [76]. The bacteria are able to degrade quorum-sensing signal molecules produced by *Pectobacterium* spp. and *Dickeya* spp., which is a useful and effective strategy for the control of the bacteria by preventing the secretion of large quantities of pectolytic enzymes to macerate tuber tissues [77].

8. Bacterial wilt of potato

Bacterial wilt caused by *R. solanacearum* (Smith) Yabuuchi *et al.* is one of the most important and destructive bacterial diseases, widely distributed in tropical, subtropical and some warm temperate regions of the world [78]. This disease affects the potato crop in 3.75 million acres in about 80 countries with global damage estimate exceeding $ 950 million per year. It damages

http://dx.doi.org/10.5772/64853

the crop in two ways: first way, premature wilting and death of plants and second way, causing rot of tubers in storage and transit [79]. The pathogen is soil and seed borne in nature. Bacterial wilt has become a limiting factor in potato cultivation that may cause yield loss to the tune of 30–70 % in India [80]. Avirulent strains of *R. solanacearum*, *Pseudomonas* spp., *Bacillus* spp. and *Streptomyces* spp. are well-known biocontrol agents (BCAs). New or uncommon BCAs have also been identified, such as *Acinetobacter* sp., *Burkholderia* sp. and *Paenibacillus* sp. [81]. Vesicular-arbuscular mycorrhizae (VAM) is known to reduce disease incidence and enhance plant growth. The potential of vesicular-arbuscular mycorrhizae was evaluated for protection of plants from bacterial wilt in the Philippines; VAM increased growth and yield of tomatoes and reduced infection by *R. solanacearum*. This may be due to competition or the mechanical barrier in the form of VAM vesicles and hyphae that inhibit the bacterial pathogen from deeper penetration into host tissues [82]. Treatment of tubers with avirulent strain of *R. solanacearum* and strain of *P. fluorescens* caused a significant reduction in disease severity of bacterial wilt of potato [83].

9. Conclusion

Different bio-agents including fungal and bacterial were reported by various researchers for management of potato diseases. Efficacy of bio-agents is varied from lab to field conditions. It might be due to non-synchrony environment between lab and field. Some *Trichoderma* spp, *Pseudomonas* spp and *Bacillus* spp exhibited significant result to reduce the incidence of potato diseases under both lab and field. These bio-agents must be applied at larger scale. Moreover, new bio-agents with a wider range of adoptability still require to be explored. A bio-agent should be applied for specific disease where it performs highest efficacy and in particular regions. It is the important constituent of organic potato production system.

Author details

Mehi Lal[1*], Saurabh Yadav[1], Vivek Singh[2] and M Nagesh[3]

*Address all correspondence to: mehilalonline@gmail.com

1 ICAR-Central Potato Research Institute Campus, Modipuram, Uttar Pradesh, India

2 Department of Plant Pathology, Banda University of Agriculture & Technology Banda, Uttar Pradesh, India

3 ICAR-Central Potato Research Institute, Shimla, Himachal Pradesh, India

References

[1] Baker KF, Cook RJ. Biological control of plant pathogens. W H Freeman and Co, San Francisco, California, 1974. 433 p (Book, reprinted in 1982, The American Phytopathological Society, St Paul, Minnesota).

[2] Chaube HS, Pundhir VS. Crop diseases and their management. Prentice Hall of India Private Limited, New Delhi, 2005. 703 p.

[3] Hoitink HAJ, Krause MS, Han DY. Spectrum and mechanisms of plant disease control with composts. In: Peter JS, Brian AK (eds). Compost utilization in horticultural cropping systems. Lewis Publishers, Boca Raton, FL, 2001. 263 p.

[4] Bharath BG, Lokesh S, Shetty HS. Effects of fungicides and bioagents on seed mycoflora, growth and yield of watermelon. Integr Biol Sci. 2005;9:75-78.

[5] Glare T, Caradus J, Gelernter W, Jackson T, Keyhani N, Kohl J, Marrone P, Morin L, Stewart A. Have biopesticides come of age? Trends Biotechnol. 2012;30:250-258.

[6] Weindling R. Trichoderma lignorum as a parasite of other soil fungi. Phytopathology. 1932;22: 838-845.

[7] Brimner TA, Boland GJ. A review of the non-target effects of fungi used to biologically control plant diseases. Agric Ecosyst Environ. 2003;100:3-16.

[8] Louche-Tessandier D, Samson G, Hernandez-Sebastia C, Chagvardieff P, Desjardins Y. Importance of light and CO_2 on the effects of endomycorrhizal colonization on growth and photosynthesis of potato plantlets (Solanum tuberosum) in an in vitro tripartite system. New Phytol. 1999;142:539-550.

[9] Alexopoulos CJ, Mims CW, Blackwell M. Introductory mycology, 4th ed. John Wiley & Sons. Inc, USA, 1996.

[10] Large EC. The advance of the fungi. Jonathan Cape, London, 1940. 488 p.

[11] Butler EJ. Fungi and disease in Plants. Thacker Spink and Co., Calcutta, 1918.

[12] Butler K. Potato disease of India. Agric Ledger Crop Dis Pest Ser. 1903;8:87-124.

[13] Dastur, JF. Conditions influencing the distribution of potato blight in India. Agric J India (Special Indian Congress). 1917;12:90-95.

[14] Lal M, Arora R K, Maheshwari Uma, Rawal S and Yadav S. Impact of late blight occurrence on potato productivity during 2013-2014. Int J Agric Stat Sci. 2016;12 (1): 187-192.

[15] Fry WE, Birch PRJ, Judelson HS, Grünwald NJ, Danies G, Everts KL, Gevens AJ, Gugino BK, Johnson DA, Johnson SB, McGrath MT, Myers KL, Ristaino JB, Roberts PD, Secor G, Smart CD. Five reasons to consider Phytophthora infestans a re-emerging pathogen. Phytopathology. 2015;105:966-981.

[16] Sas-Piotrowska, Piotrowski B, Misiak WM. The growth and development of potato pathogens on media with extracts from Polygonaceae plants. I. Pathogens causing dry leaf-spot disease. Phytopathol Polonica. 1996;11:103-109.

[17] Blaser P, Steiner U, Lyr H, Russel PE, Dehne HW, Sisler HD. Antifungal activity of plant extracts against potato late blight (Phytophthora infestans). Modern fungicides and antifungal compounds II Friedrichroda, Thuringia, Germany 24–29 May 1998. 1999. 491-499 pp

[18] Paik SB. Screening for antagonistic plants for control of Phytophthora spp. in soil. Korean J Mycol. 1989;17:39-47

[19] Khan MA, Singh BP, Kaushik SK, Lal M. Evaluation of antifungal potential extracts against potato pathogens. National Seminar on Emerging Problems of Potato (1–2 November 2014, CPRI, Shimla), 2014. 143 p.

[20] Ajay S, Sunaina V. Direct inhibition of Phytophthora infestans, the causal organism of late blight of potato by Bacillus antagonist. Potato J. 2005;32:179-180.

[21] Lal M, Yadav S, Kaushik SK, Sharma S, Chand S. Integrated management of late blight of potato. National Seminar on Emerging Problems of Potato (1–2 November 2014, CPRI, Shimla), 2014. 199 p.

[22] Indian Institute of Spices Research. PhytoFuRa Annual Report 2013–2014. IISR, Kozhikode, 2014:20-26.

[23] Ramos L, Ciampi L, Gonzales S. Biological control of Phytophthora infestans in potato plants. Simiente. 1993; 63:53-54.

[24] Sanchez V, Bustamante E, Shattock R. Selection of antagonists for biological control of Phytophthora infestans in tomato. Manejo Integrado de Plagas. 1998;48:25-34.

[25] Sadlers HM. Use of bacteria in controlling fungal diseases. Gemuse Munchen. 1996;32: 180-181.

[26] Garita VS, Bustamante E, Shattock R. Microbiological control of Phytophthora infestans on tomato. Manejo Integrado de Plagas.1999; 51:47-58.

[27] Stephan D, Koch E, Elad Y, Kohl J, Shtienberg D. Screening of plant extracts, micro-organisms and commercial preparations for biocontrol of Phytophthora infestans on detached potato leaves. Bulletin OILB-SROP. 2002;25:391-394.

[28] Gupta H, Singh BP, Mohan J. Biocontrol of late blight of potato. Potato J. 2004;31:39-42.

[29] CPRI. 1999. Annual Progress Report. Central Potato Research Institute, Shimla, India.

[30] Roy S, Singh BP, Bhattacharyya SK. Biocontrol of late blight of potato. Phytophthora Newslett. 1991;17:18.

[31] Lal M, Singh AP, Tomar S, Hussain T, Sharma S, Kaushik SK, Singh BP. Antagonistic effect of bioagents against three potato fungal diseases and their fungicidal sensitivity. Vegetos. 2013; 26:362-367.

[32] Kim HY, Choi GJ, Lee HB, Lee SW, Lim HK, Jang KS, Son SW, Lee SO, Cho KY, Sung ND, Kim JC. Some fungal endophytes from vegetable crops and their anti-oomycete activities against tomato late blight. Lett Appl Microbiol. 2007;44:332-337.

[33] Chen SY, Wei YH, Chang JS. Repeated pH-satisfied batch fermentation for rhamnolipid production with indigenous Pseudomonas aeruginosa S2. Appl Micro Biotechnol. 2007;76:67-74.

[34] Tomar S, Singh BP, Khan MA, Kumar S, Sharma S, Lal M. Identification of Pseudomonas aeruginosa strain producing biosurfactant with antifungal activity against Phytophthora infestans. Potato J. 2013;40:155-163.

[35] Tomar S, Singh BP, Lal M, Khan MA, Hussain T, Sharma S, Kaushik SK, Kumar S. Screening of novel microorganisms for biosurfactant and biocontrol activity against Phytophthora infestans. J Environ Biol. 2014;35:893-899.

[36] Tomar S, Singh BP. Microbial origin, classification and application of biosurfactant. Indian J Anim Nutr. 2014;7:2060-2069.

[37] Banat IM, Makkar RS, Cameotra SS. Potential commercial application of microbial surfactants. Appl Microbiol Biotechnol. 2000;53:495-508.

[38] Shuman JL. Integrating a host resistance factor into a potato early blight-forecasting model. M.Sc. thesis, Pennsylvania State University, 1995.

[39] Nashwa SMA, Abo-Elyousr KAM. Evaluation of various plant extracts against the early blight disease of tomato plants under greenhouse and field conditions. Plant Protect Sci. 2012;48:74-79.

[40] Yadav R, Pathak SP. Management of early blight of potato through fungicides and botanicals and bioagents. Plant. 2011;11:1143-1145.

[41] Mane MM, Lal AA, Zghair NQ, Sobita S. Efficacy of certain bioagents and fungicides against early blight of potato (Solanum tuberosum L.). Int J Plant Protect. 2014;7:433-436

[42] Lal M, Sharma S, Yadav S, Kaushik SK. Bioefficacy of new molecule: penflufen 240 fs against black scurf of potato. Int J Agric Stat Sci. 2014;10 (supplement 1):63-66.

[43] Arora RK. Management of black scurf of potato with the integrated use of Trichoderma viride and boric acid. Potato J. 2008;35:130-133.

[44] Mohsin T, Yasmin S, Hafeez FY. Biological control of potato black scurf by rhizosphere associated bacteria. Braz J Microbiol. 2010; 41:439-451.

[45] Basu A. Employing eco-friendly potato disease management allows organic tropical Indian production systems to prosper. Asian J Food Agro-Indust. Special Issue. 2009;S80-S87.

[46] Selva SK, Rao MRK, Kumar RD, Panwar S, Prasad CS. Biocontrol by plant growth promoting rhizobacteria against black scurf and stem canker disease of potato caused by Rhizoctonia solani. Arch Phytopathol Plant Protect. 2013;46:4487-4502.

[47] Arora, RK, Somani AK. A bioformulation for control of soil and tuber borne diseases of potato. J Indian Potato Assoc. 2001;28:88-89.

[48] Prasad CS, Gupta V. Studies on bio efficacy of Trichoderma harzianum in management of stem rot of potato (Rhizoctonia solani). Natl Acad Sci Lett. 2002;25:357-359.

[49] Hazarika DK, Phookan AK, Das KK, Dubey LN, Das BC. Biological management of black scurf of potato. In Paul Khurana SM, Shekhawat GS, Singh BP and Pandey SK (eds). Potato global research and development. Proceedings (Volume I) of Global Conference on Potato, New Delhi, December 6–11, 1999. Indian Potato Association. 2000; 401-404

[50] Bhattacharyya, SK, Bahal VK, Bist BS. Effect of crop rotation on potato black scurf incidence. J Indian Potato Assoc. 1977;4:1-4.

[51] Larkin RP, Griffin, TS. Control of soilborne potato diseases using brassica green manures. Crop Protect J. 2007; 26:1067-1077.

[52] Garbeva P, van Elsas JD, van Veen JA. Rhizosphere microbial community and its response to plant species and soil history. Plant Soil. 2008; 302:19-32.

[53] Hussain A, Awan MS, Khan SW, Ali MAS, Ali QAA. Bioefficacy of botanical extracts and bioagents against sclerotial isolates of Rhizoctonia solani. J Biodivers Environ Sci. 2014; 4:370-380.

[54] Sharma S, Lal M. Dry rot. In BP Singh, M Nagesh, Sanjeev Sharma, Vinay Sagar, A Jeevvlatha and J Sridhar (eds). A manual on diseases and pest of potato-Technical Bulletin No. 101. ICAR-Central Potato Research Institute, Shimla, HP, India. 2015. p. 17-19.

[55] Sundaramoorthy S, Balabaskar P. Biocontrol efficacy of Trichoderma spp. against wilt of tomato caused by Fusarium oxysporum f. sp. lycopersici. J Appl Biol Biotechnol. 2013; 1(03):036-040.

[56] Singh BP, Nagaich BB, Saxena SK. Studies on the effect of organic amendments on fusarium wilt of potato. J Indian Potato Assoc. 1988; 15:60-67.

[57] Abeer H, Makhlouf, Rehab A. Biological and nanocomposite control of fusarium wilt of potato caused by Fusarium Oxysporum f. sp. Tuberosi G.J.B.A.H.S. 2015;4:151-163.

[58] El-Shennawy MZ, EZ, Khalifa MM, Ammar EM, Mousa, Hafez, SL. Biological control of the disease complex on potato caused by root-knot nematode and Fusarium wilt fungus. Nematol Medit. 2012;40:169-172.

[59] Bouchek-Mechiche K, Pasco C, Andrivon D, Jouan B. Differences in host range, pathogenicity to potato cultivars and response to soil temperature among Streptomyces species causing common and netted scab in France. Plant Pathol. 2000;49:3-10.

[60] Daqun L, Neil A, Anderson A, Kinkel L. Biological control of potato scab in the field with antagonistic Streptomyces scabies. Phytopathol. 1995;85:827-831.

[61] Lorang JM, Liu D, Anderson NA, Schottel JL. Identification of potato scab inducing and suppressive species of Streptomyces. Phytopathol. 1995;85:261-268.

[62] Rafiq M, Bukhari A. Evaluation of different antagonistic fungi against common scab of potato. Mycopath. 2014;12:63-67.

[63] Singhai PK, Sarma BK, Srivastava JS. Biological management of common scab of potato through Pseudomonas species and vermicompost. Bio Cont. 2011;57:150-157.

[64] Tsuyoshi T, Nishi Y, Mori K, Shirao T, Iida T, Uzuhashi S, Ohkuma M, Ikeda S. Rice bran amendment suppresses potato common scab by increasing antagonistic bacterial community levels in the rhizosphere. Phytopathol. 2016;106(7):719-728.

[65] Tanaka H, Negishi H, Maeda H. Control of tobacco bacterial wilt by an avirulent strain of Pseudomonas solanacearum M4S and its bacteriophage. Ann Phytopathol Soc Japan. 1990;56:243-246.

[66] McKenna F, El-Tarabily KA, Hardy GESTJ, Dell B. Novel in vivo use of polyvalent Streptomyces phage to disinfest Streptomyces scabies-infected seed potatoes. Plant Pathol. 2001;50:666-675.

[67] Han JS, Cheng JH, Yoon TM, Song J, Rajkarnikar A, Kim WG, Yoo ID, Yang YY, Suh JW. Biological control agent of common scab disease by antagonistic strain Bacillus sp. sunhua. J Appl Microbiol. 2005;99:213-221.

[68] Liu DQ, Anderson NA, Kinkel LL. Selection and characterization of strains of Streptomyces suppressive to the potato scab pathogen. Can J Microbiol. 1996;42:487-502.

[69] Liu D, Anderson NA, Kinkel LL. Biological control of potato scab in the field with antagonistic Streptomyces scabies. Phytopathology. 1995;85:827-831.

[70] Pe'rombelon MCM. Potato diseases caused by soft rot erwinias: an overview of pathogenesis. Plant Pathol. 2002;51:1-12.

[71] Ansermet M, Schaerer S, Kellenberger I, Tallant M, Dupuis B. Influence of seed-borne and soil-carried inocula of Dickeya spp. on potato plant transpiration and symptom expression. Eur J Plant Pathol. 2016;145:459-467.

[72] De Weger LA, van Boxtel R, van der Burg B, Gruters RA, Geels FP, Schippers B, Lugtenberg BJJ. Siderophores and outer membrane proteins of antagonistic, plant growth-stimulating root-colonizing Pseudomonas spp. J Bacteriol. 1986;165:585-594.

[73] Cronin D, Moënne-Loccoz Y, Fenton A, Dunne C, Dowling DN, O'Gara F. Ecological interaction of a biocontrol Pseudomonas fluorescens strain producing 2,4-diacetyl-phloroglucinol with the soft rot potato pathogen Erwinia carotovora subsp. atroseptica. FEMS Microbiol Ecol. 1997;23:95-106.

[74] Sharga BM, Lyon GD. Bacillus subtilis BS 107 as an antagonist of potato blackleg and soft rot bacteria. Can J Microbiol. 1998;44:777-783.

[75] Cladera-Olivera F, Caron GR, Motta AS, Souto AA, Brandelli A. Bacteriocin-like substance inhibits potato soft rot caused by Erwinia carotovora. Can J Microbiol. 2006;52:533-539.

[76] Kastelein P, Schepel E, Mulder A, Turkensteen L, Van Vuurde J. Preliminary selection of antagonists of Erwinia carotovora subsp. atroseptica (Van Hall) dye for application during green crop lifting of seed potato tubers. Potato Res. 1999;42:161-71.

[77] Jafra S, Przysowa J, Czajkowski R, Michta A, Garbeva P, van der Wolf JM. Detection and characterization of bacteria from the potato rhizosphere degrading N-acyl-homoserine lactone. Can J Microbiol. 2006;52:1006-1015.

[78] Zhou XZ, Zhang QW, Liu XX. Effects of agricultural streptomycin and rhizobacteria Bs 8093 on soil microbial communities estimated by analysis of phospholipid fatty acids. Indian J Agric Sci. 2011; 80:42-50.

[79] Sagar V. Bacterial wilt and brown rot. In BP singh, M Nagesh, Sanjeev Sharma, Vinay Sagar, A Jeevvlatha and J Sridhar(eds). A manual on diseases and pest of potato-Technical Bulletin No. 101 ICAR-central potato research institute, Shimla, HP, India. 2015. p. 20-23.

[80] Somani AK, Chakrabarti SK, Pandey SK. Spread of bacterial wilt and brown rot of potato in Indore region of Madhya Pradesh. CPRI News Letter no. 42 (June), 2010. 16-17.

[81] Yuliar, Nion YA, and Toyota K. Recent trends in control methods for bacterial wilt diseases caused by Ralstonia solanacearum. Microbe Environ. 2015;30:1-11.

[82] Halos PM, Zorilla RA. Vesicular-arbuscular mycorrhizae increase growth and yield of tomatoes and reduce infection by Pseudomonas solanacearum. Philipp. Agric. 1979;62:309-315.

[83] Kempe J, Sequeira L. Biological control of bacterial wilt of potatoes: attempts to induce resistance by treating tubers with bacteria. Plant Dis. 1983;67:499-503.

Chapter 2

Plant Health

Munazza Gull

Additional information is available at the end of the chapter

http://dx.doi.org/10.5772/64869

Abstract

Environment friendly control of plant disease is an emerging need for agriculture in the twenty-first century. Biological control using antimicrobial producing rhizobacteria to suppress plant diseases and promote plant health offers a powerful alternative to the use of synthetic chemicals. Many studies have been conducted to identify the specific traits by which plant growth-promoting rhizobacteria (PGPR) promote plant growth. Most of these studies were limited to examining just one or two of these traits. The plant growth-promoting rhizobacteria produce a wide variety of antimicrobial compounds against pathogens. The addition of antagonistic antimicrobial producing bacterial strains, either individual or as mixture in combination with fungicide, significantly decreased the plant disease stress. A single PGPR strain can produce different kinds of antimicrobial defense compounds to compete pathogens. A biocontrol agent possessing multimechanism systems of defense can antagonize pathogens in a better way. This research chapter highlights the current advancements about plant-PGPR interactions focusing on the principles and defensive mechanisms of PGPR during disease stress conditions and their potential use for the biocontrol of plant diseases. The integrated use of genetic, molecular, and ecological approaches will form the basis for significant future advances in biocontrol research against plant diseases.

Keywords: plant health, plant growth-promoting bacteria, microbial pathogens, antimicrobial compounds, biocontrol

1. Introduction

Pathogenic microorganisms affecting plant health are a major and chronic threat to food production and ecosystem stability worldwide. Agricultural yield and production increased in past few decades due to intensive use of agrochemicals providing more stable and reliable method for crop protection. The increasing use of these fertilizers and pesticides results in several negative effects on the environment, i.e., development of pathogen resistance and adverse impacts on nontarget organisms. In addition, the high cost of these fertilizers and pesticides and increasing demand of consumers for chemical-free food have led to a search for alternative natural products. There are many plant diseases for which chemical pesticides and stable protection from pathogens are not available. In this scenario, an alternative way of reducing the use of agrochemicals in agriculture, which also provides an effective disease protection and continuous supply of natural food, is biological control [1].

The ability of microorganisms to respond to stress in their environment is the key to their survival. In general terms, any condition that prevents an organism from growing at its optimal rate may be considered a form of environmental stress. For an organism to survive, it must respond to the environmental conditions imposed upon it, whether it is the absence of a nutrient, extremes in temperature, pH or oxidative state, or the presence of toxic compounds. Bacterial responses to these factors are varied and can include the expression of new proteins, the loss of plasmids, changes in membrane fatty acid content, changes in DNA super coiling and, in some cases, cross-tolerances to yet unencountered forms of environmental stress [2].

The lack of homogeneity and varied make up of soil dictates that organisms living in it must be able to adapt and survive. It was the purpose of this study to examine the interplay of nutrient limitation, specifically iron, and the presence of a wide array of antimicrobial compounds on the ability of the plant growth-promoting rhizobacteria to adapt to its environment and suppress the pathogenic disease. To understand the role of antimicrobial compounds in biocontrol of soil-borne pathogens, an overview of the plant rhizospheric ecology, PGPR, and biocontrol mechanisms is first required.

1.1. Plant rhizosphere

The "rhizosphere" can be defined as the part of soil around plant roots where bacterial growth is stimulated. It is the habitat where several biologically important processes and plant microbe interactions take place. A diverse range of microorganisms are populated in rhizosphere and the bacteria colonizing this habitat are usually named as rhizobacteria.

1.1.1. Plant growth-promoting rhizobacteria (PGPR)

There has been a large body of literature describing potential uses of plant-associated bacteria as agents stimulating plant growth and managing soil and plant health. Plant growth-promoting bacteria (PGPR) are associated with almost all plant species in a range of environments. Plant growth-promoting rhizobacteria (PGPR) colonizing the root surfaces and the closely adhering soil interface are the extensively and widely studied group. These PGPR can also enter into the interior parts of roots and establish populations of endophytic bacteria. Majority of these rhizobacteria transcend the barrier of endodermis, penetrating from the cortex of root to the vascular system, and finally reach in the upper parts of plants like stem, leaves, and tubers [3]. The ability of bacteria to selectively adapt these specific ecological niches depends on the extent of endophytic colonization of host plant organs and tissues. Conse-

http://dx.doi.org/10.5772/64869

quently, without harming the plant, eco-friendly associations between bacteria and host plants become established.

It is generally considered that many endophytic bacteria are the final product of a plant microbe process of colonization occurred in the root zone [4].

1.1.2. Direct plant growth promotion

PGPR can influence plant growth directly. These ways differ species to species and even from one bacterial strain to other strain. Rhizobia as symbiotic plant colonizers contribute to plant growth stimulation by enhancing nitrogen fixation. Free-living rhizobacteria usually do not depend on single plant growth-promoting mechanism. Several PGPR are also able to provide the plant with sufficient iron in iron-limiting soils or other important minerals, e.g., phosphate and zinc [5].

1.1.3. Indirect plant growth promotion

Indirect growth promotion occurs when PGPR promote plant growth by improving growth-restricting conditions. This can happen directly by producing antagonistic substances or indirectly by inducing resistance in host plants to a broad spectrum of pathogens. A bacterium can affect plant growth by one or more of these mechanisms and also use different abilities for growth promotion at various times during the life cycle of the plants. The widely recognized mechanisms of biocontrol mediated by PGPR are competent for an ecological niche or a substrate, production of inhibitory allelochemicals, induction of systemic resistance (ISR), and/or abiotic stresses [6].

1.1.4. Competitive root colonization

Successful application of PGPR has been hampered by inconsistent performance under field conditions. This is usually due to their poor and unstable rhizosphere competence. Effective root colonization with the ability to survive and proliferate along growing plant roots for a definitive time period in the presence of the other indigenous microflora results in effective rhizosphere competence development. Rhizosphere competence is considered as a prerequisite of effective biological control. Understanding root-microbe interactions as affected by genetic and environmental factors in spatial temporal contexts could significantly contribute to improve the efficacy of these biocontrol agents under wide range of field conditions [7]. Successful and stable application of PGPR is most directly affected by competition for root niches and bacterial determinants.

Root exudates determine which microorganism colonizes roots in the rhizosphere. It is now known that plant roots also generate electrical signals and zoospores of oomycetic pathogens take advantage of these signals to guide their movements toward the root surface. Both physical and chemical benefits to plants are provided by exudates, e.g., reduce the friction between root tips and the soil by root mucilages and reduction of root desiccation establish the effective contact between the root tips and the soil and contribute to soil structural stability. Root exudates also attract microorganism. Conversely, rhizobacteria can also elicit root exudation in a specific manner, e.g., metabolites produced by *Pseudomonas aeruginosa* stimulate root exudates by perennial ryegrass 12-fold [8]. Root exudates can also be used as effective and stable antimicrobial agents, which can provide ecological niche an advantage to organisms that have perfect enzymatic mechanisms to detoxify them. Genetic and environmental conditions control the quantity and composition of chemoattractants and antimicrobials produced by plant roots. This indicates that PGPR competence is highly affected by the ability of rhizobacteria to survive under specific environment and to adapt the changing conditions rapidly [9].

Important bacterial traits identified for effective and stable root colonization are linked to phase variation, a regulatory process for DNA rearrangements controlled by site-specific recombinase enzyme. In some PGPR, efficient root colonization is subject to their ability to secrete an effective site-specific recombinase. This importance has been found when a site-specific recombinase gene from a rhizosphere-competent *P. fluorescens* was transferred into a rhizosphere-incompetent *Pseudomonas* strain and it enhanced its ability of root tips colonization [10].

2. Biocontrol of soil pathogens by antimicrobial producing rhizobacteria

A great diversity of rhizospheric microorganisms has been studied, characterized, and analyzed as biocontrol agents against many soil-borne pathogens over the past decades. Such microorganisms can produce substances that may reduce the damage caused by phytopathogens, e.g., by producing antibiotics, siderophores, and variety of enzymes. These microorganisms can also serve as competitors of pathogens for root colonization sites and nutrients. Biocontrol has not yet become widely popular and applied as alternative source of agrochemicals due to several factors. For example, the efficiency and activity of a biocontrol strain under field condition is usually affected by changing environmental conditions: water contents, pH, temperature, and interactions with other microorganisms. As a result, these biocontrol agents that showed promising plant growth stimulation and disease protection traits in initial laboratory experiments failed to be efficient rhizosphere colonizers under more complex biological field conditions. This highlights the need to address these limitations by extensive study of genetic, biochemical, and physiological factors that contribute to the effective and successful activity of biocontrol agents under wide range of environmental conditions.

2.1. Antibiosis

Antibiotics play a very important role in pant disease suppression by biocontrol agaents. Molecular and genetic tools could be effective in this regard because mutant defective in antibiotic production are easily obtained and studied by *in vitro* assays. With respect to the production of antibiotics, the most widely studied group of rhizosphere bacteria is fluorescent pseudomonads. Phenazine derivatives produce by fluorescent pseudomonads were the first biocontrol antibiotics described. Transposon insertion mutations elucidated their role that results in a defective and insufficient production of phenazine-1-carboxylate. As a result, disease suppressive activity has been reduced in these mutants [11]. The functional genes encoding the metabolic synthesis enzymes had been isolated, identified, and their up- and

http://dx.doi.org/10.5772/64869

down-regulation were studied. The presence of populations of other bacteria can influence phenazine production by *P. aureofaciens*, since mutants lacking the ability to produce, and autoinducer signal required for induction of antibiotic synthesis can use autoinducers produced by other (related) rhizosphere inhabitants. Also, other environmental sensors such as regulatory proteins *Gacy* and *ApdA* can influence the production of secondary metabolites involved in Pseudomonad biocontrol [12]. In addition, sigma factors are important for regulation of antibiotic production in fluorescent pseudomonads; housekeeping factor sigma 70 and the stress-related sigma have critical roles in production of antibiotic metabolites in disease suppression.

Antibiosis as a biocontrol mechanism of PGPR has become increasingly popular, better studied and used over the past decades. A large variety of antibiotics have been identified and formulated such as amphisin, 2,4-diacetylphloroglucinol (DAPG), oomycin A, phenazine, pyrrolnitrin, pyoluteorin, tensin, tropolone, hydrogen cyanide, and cyclic lipopeptides produced by *Pseudomonas* spp., and kanosamine oligomycin A, zwittermicin A, and xanthobaccin produced by *Bacillus*, *Streptomyces*, and *Stenotrophomonas* spp. [13]. Some antibiotics produced by PGPR are finding new pharmaceutical uses and these rhizobacteria opened an untapped and continuous resource for compounds to deal with the alarming arouse of multidrug-resistant pathogenic bacteria in human.

Regulatory cascades of these efficient antibiotics include *GacA/GacS* or *GrrA/GrrS*, *RpoD*, *RpoS*, and N-acyl homoserine lactone derivatives [14] and positive autoregulation [15]. Antibiotic synthesis is tightly linked and associated to the overall metabolism of the cell. Metabolic regulation of cell is dictated by nutrient availability and other environmental stimuli, such as pH, temperature, water, major and minor minerals, type of carbon source and supply, and other variety of parameters [16]. Genetic stability/instability of bacteria, affecting their ability to produce secondary metabolites, has been influenced by trace elements particularly zinc and carbon source levels. It is interesting to found that many bacterial strains produce pallet of secondary antimicrobial metabolites and the conditions favoring the production of one compound may not favor another metabolite mechanism. This wide variety of biocontrol strains may enable antagonistic bacteria to suppress the pathogens under the widest range of environmental conditions effectively and with stability. For example, in *P. fluorescens* CHA0 biosynthesis of diacetylphloroglucinol (DAPG) is stimulated, and pyoluteorin is repressed if glucose present and used as a carbon source. As glucose is depleted and its concentration decreased, pyoluteorin becomes the more abundant antimicrobial compound produced by this strain. This provides a kind of stability and flexibility as well for the antagonistic bacteria when dealt with a different or a changeable environment. Antibiotic biosynthesis can also be influenced by biotic conditions [17]. For example, bacterial metabolites pyoluteorin and salicylates can increase or decrease DAPG production by *P. fluorescens* CHA0. In addition, plant growth and development also influence antibiotic production because biological activity of DAPG-producing bacteria is not initiated by the root exudates of young plants but is induced by the root exudates of older plants, which gives in a strong selective pressure against other rhizosphere microorganisms sharing same ecological niche. Plant host genotype and their regulation also play a significant role in the disease-suppressive interaction of plant with a microbial biocontrol agent [18].

2.2. Hydrolytic enzymes production

A variety of microorganisms also shows hyperparasitic mechanism, attacking plant pathogens by excreting cell wall enzymes called hydrolases. *Streptomyces plymuthica* C48 produced chitinase, which inhibited germination of spores and germ tube elongation in *Botrytis cinerea* effectively. The production of extracellular chitinase is considered a strong defensive mechanism for *Serratia marcescens* to act as antagonistic organism against *Sclerotium rolfsii* and for *Paenibacillus* sp. strain 300 and *Streptomyces* sp. strain 385 to suppress *Fusarium oxysporum* f. sp. *cucumerinum*. It has been also studied that extracellular chitinase and laminarinase produced by *Pseudomonas stutzeri* break and lyse mycelia of *F. solani* [19]. Chitinolytic activity found less utilized defensive mechanism in PGPB such as *S. plymutica* IC14 when used against *S. sclerotiorum* and *B. cinerea*, synthesis of proteases and other biocontrol characters are involved [20]. *B. cepacia* produces ß-1,3-glucanase that destroys the cell wall integrity of fungal strains *R. solani, S. rolfsii,* and *Pythium ultimum*. Production of lytic enzymes (proteases and chitinases, in particular) is regulated by *GacA/GacS* or *GrrA/GrrS* regulatory system and colony phase variation [21].

2.3. Detoxification and degradation of virulence factors

Biological control exhibits antagonism by detoxification of pathogen virulence factors also. For example, few biocontrol microorganisms are capable of detoxifying albicidin toxin synthesized by *Xanthomonas albilineans*. The detoxification mechanisms involve production of a protein that binds to the toxin reversibly in both *Klebsiella oxytoca* and *Alcaligenes denitrificans*, as well as an irreversible detoxification of albicidin mediated by an esterase enzyme found in *Pantoea dispersa* [22]. Different strains of *B. cepacia* and *Ralstonia solanacearum* can also lyse phytotoxin fusaric acid produced by different *Fusarium* species. Mostly pathogen toxins exhibit a broad spectrum activity of defense mechanisms and can restrain growth of microbial competitors. They can detoxify antibiotics produced by some biocontrol microorganisms as a self-defense mechanism against biocontrol agents [23].

It has been discovered recently that few PGPB show pathogen quorum sensing ability by degrading autoinducer signals, thereby blocking expression of various virulence genes. Bacterial plant pathogens use autoinducer-mediated quorum sensing to switch on gene cascades for their key virulence factors (e.g., cell-degrading enzymes and phytotoxin production). This approach holds tremendous antagonistic potential for suppression of diseases, even after the onset of infection effectively.

Biocontrol activity of microorgansims by production of allelochemicals has been studied widely with free-living rhizobacteria. Similar antagonistic mechanisms are used by endophytic bacteria as they can also synthesize antagonistic metabolites against plant pathogens. For example, it has been established that antibiotics munumbicins produced by the endophytic bacteria *Streptomyces* sp. strain NRRL 30562 isolated from *Kennedia nigriscans* can suppress *in vitro* growth of phytopathogenic fungi, *P. ultimum*, and *F. oxysporum*, effectively.

Certain endophytic bacteria isolated from field-grown potato plants can suppress the *in vitro* growth of *Streptomyces scabies* and *Xanthomonas campestris* through production of siderophores, antibiotics, and other antagonistic metabolites [24]. The ability to inhibit pathogenic growth by endophytic bacteria isolated from potato tubers decreases as the bacteria colonize the host plant's interior suggesting that bacterial adaptation to this habitat occurs within their host and may be tissue type and tissue site specific. It has been found that the endophytic bacterial strain *P. fluorescens* FPT 9601 can produce DAPG and deposit DAPG crystals around and in the roots of tomato. This ability of endophytic bacteria to produce antibiotics in plants is very promising and could be used as antagonistic mechanism against pathogens [25].

2.4. Induction of systemic resistance

An advanced level of resistance at sites within that plant distant to those parts where infection had occurred is called systemic resistance. PGPR-triggered ISR provides strength and integrity to plant cell walls and boost host physiological and metabolic responses, leading to an increased production of plant defense chemicals against plant pathogens or abiotic stress factors. This recognition mediates the extracellular to intracellular signals. Then, the metabolite by itself or a signal generated by the plant cell turns on a signal transduction cascade. Consequently, distant plant cells, triggering the activation of defense arsenal of the diseased host plant, recognize the translocated signals. The pathways of signal transduction are activated upon microbial challenge, which results in activation of different sets of effector molecules.

Salicylic acid (SA), jasmonate (JA), and ethylene (ET) are the signaling molecules when accumulating trigger the defense responses and, if used exogenously, are even sufficient to induce resistance and suppress disease [26]. These SA signaling molecules activate genes encoding pathogenesis-related proteins (PRs). These self-defense proteins have antimicrobial potential. ET is involved in the regulation and expression of the defensive genes encoding *Hel* (a hevein-like protein basic chitinase (*ChiB*) and a plant defensin (*Pdf1.2*)) [27]. JA has been found to activate and regulate the genes encoding these three proteins. They possess antifungal activity. Furthermore, JA also activates the gene encoding a vegetative storage protein, *Atvsp*. These proteins accumulate in vacuoles but their potent role in antagonistic mechanism has not yet been confirmed.

Two defense pathways, induced systemic resistance and systemic acquired resistance (SAR), are found induced in *Arabidopsis*. ISR is a bacterial-mediated systemic resistance that causes no damage to plant but SAR is induced by foliar pathogens and results in activation of resistance mechanisms in uninfected parts of plant. It is established that in SAR, a first infection predisposes the plant to resist further attacks of pathogens. SAR mediation relies on the accumulation of SA and requires the regulatory inducer protein NPR1. In addition to SA accumulation, several JA- and ET-dependent resistance defense mechanisms that are independent of SA have also been described [28]. JA and ET act synergistically in inducing cascade of genes for numerous PR proteins. ET has been found to enhance JA-dependent resistant responses but SA suppresses the JA-dependent defense gene expression. JA has also been reported to interface with SA-dependent defense signaling mechanism. ISR can be induced in plants that are not capable to accumulate SA (*NahG* mutant plants). This shows that SA is not required for ISR induction in *Arabidopsis*. PR proteins do not found accumulated in induced plants. However, the regulator NPR1 protein is required for expression of ISR [29]. ET- or JA-responsive defective genes *etrl, ein2, ein7,* or *jar1* in *Arabidopsis* mutant plants conferring a decreased sensitivity to ET and JA. They also found defective in their expression of ISR. JA application to wild-type plants induces a defense resistance that is not linked with the accumulation of PRs but is dependent on a functional *npr1* gene. These results showed that response to JA and ET is sequentially required in the ISR signal transduction pathway. ISR-mediated defense mechanisms of PGPR varied widely among species. PsJN-grapevine interaction, a host defense reaction in *Burkholderia phytofirmans,* found associated with phenolic compound accumulation and strengthening of cell walls in the exodermis and in several cortical cell layers during endophytic bacterial colonization [30]. The type of plant response linked to antagonistic bacteria induced after pathogen infection leads to the formation of structural barriers, such as thickened cell wall and papillae due to callose deposition and phenolic compounds accumulation at the site of pathogenic attack.

2.5. Hydrogen cyanide production

Hydrogen cyanide (HCN) is released as a product of secondary metabolism by several microorganisms and affects sensitive organisms by inhibiting the synthesis of ATP mediated by cytochrome oxidase. The percentage of cyanogens found is very low among rhizobacteria [31]. Therefore, depending on the target organisms, HCN-producing microorganisms are regarded as harmful when they impair plant health and beneficial when they suppress unwanted components of a microbial community. It has been reported that an isolate capable of cyanide production could be a better biocontrol agent because cyanide acts as an inducer of plant resistance [32].

2.6. Competition for iron: Siderophores production

Siderophores, from the Greek: "iron carriers," play the role to scavenge iron from environment and to make the mineral, which is always essential, available to microbial cell. Consequently, iron becomes unavailable to microorganisms that are unable to use these siderophores and competition for iron between microorganisms seems probable. Studies of siderophore-producing microorganisms have received much attention because of the clinical application and potential utilization of these chelators in agriculture.

Fungal strains produced both extracellular and intracellular siderophores, as discovered in spores and mycelia of *Neurospora* and *Aspergillus* [33]. Whereas in marine bacteria, lipophilic siderophores have been found that do not readily diffuse into the surrounding medium except that in which vesicles are formed. This shows that environmental distribution of siderophores may vary from strain to strain. However, their general iron transport function is evident and has been analyzed by radioactive labeling experiments in a number of microorganisms.

http://dx.doi.org/10.5772/64869

However, their main function is to get iron from insoluble hydroxides or from iron adsorbed to solid surfaces. Siderophores can also extract iron by Fe (III)/ligand exchange reactions from various other soluble and insoluble iron compounds such as ferric phosphate, ferric citrate, Fe-transferrin, ferritin, or iron bound to sugars, plant flavonoid pigments and glycosides, or even from artificial chelators like EDTA and nitrilotriacetate. Therefore, even if siderophores are not involved directly in solubilization of iron, they work as carriers mediating exchange between extracellular iron storage and membrane-located siderophore transport systems of the cells.

Siderophores detection is mostly achieved in iron-limited media, which means that either a synthetic (minimal) recipe or introduction of a complexing agent will render the iron selectively unavailable. The chrome azurol sulfonate (CAS) assay has become widely used since it is comprehensive, responsive, and more convenient than other microbiological assays, which although sensitive is rigidly specific [34]. Quantitative detection of siderophores can be done by spectrophotometry and by HPLC. The presence of hydroxamate siderophores is usually detected by Csaky's test [35], and catechol siderophores are usually detected by Arnow's test [36].

Siderophores differ substantially in structure, so no uniform procedure is available for its isolation. The siderophore can be isolated as individual compound or as its iron chelate. The iron chelates has the benefit of visual color identification but the iron must be removed before any natural product can be characterized by antimicrobial assays. Complete hydrolysis in the presence of iron could damage oxidizable moieties and direct NMR analysis is ruled out by paramagnetism of the ferric ion. By a combination of NMR and mass spectroscopy, structural characterization is done in the best possible way. These methods are sensitive and capable of providing absolute answers to all arising questions. Less than half of the siderophores could be crystallized. However, by X-ray diffraction technique, the successful configuration of those molecules containing a chiral center-like siderophores could be easily possible.

Among the siderophore-producing microbes, bacteria produce both hydroxamate and catecholate siderophores but fungi produce only hydroxamate-type compounds [37].

In Gram-negative genera such as the *Enterobacteria, Pseudomonas,* nitrogen-fixing *azotobacteria,* and the plant-associated *agrobacteria,* catecholate siderophores are usually found. It has been found that lipophilicity, complex stability, high environmental pH, and a weak nitrogen metabolism may lead to the production of catecholates. *Bacillus* and *Streptomyces* Gram-positive bacteria and the ascomycetous fungi produce hydroxamate-type ferrioxamines. The basidiomycetous fungi produce ester- and peptide-containing hydroxamate siderophores mostly which are acid stable and compatible for environmental iron solubilization. Siderophore also favors the development of mycorrhizal symbiosis particularly in all terrestrial plant communities. In almost all tree species in temperate forests, ectomycorrhizal interactions typically form. Only few siderophores have been reported due to the difficulties in cultivating the pure culture of mycorrhizal fungi under iron-limited conditions. It has been reported that three mycorrhizal fungal species, *Hymenoscyphus ericae, Oidiodendron griseum,* and *Rhodothamnus chamaecistus,* an ectendomycorrhizal fungus *Wilcoxina,* and an ectomycorrhizal fungus *Cenococcusm geophilum* produce hydroxamate siderophores of the ferrichrome and fusigen class [38].

The production of siderophores has been linked to the disease suppression ability of PGPR either through a direct effect on plant by control of noxious organisms in soil or via some other routes. The involvement of siderophores in plant growth promotion and disease suppression by *Pseudomonas* strains was suggested first time. However, the first real substantiation of this concept was published by Kloepper et al. [39] who isolated the fluorescent siderophore from strain B10 and showed that it mimicked the disease suppression ability of the producing strain.

Furthermore, the inhibitory effects of both the purified siderophore and the producing strain were eliminated under high-iron conditions. Subsequent genetic evidence indicated that the inhibitory properties of certain fluorescent pseudomonads were abolished in siderophore-negative mutants. Specific siderophore-producing rhizobacteria (*Pseudomonas*) rapidly colonize plant roots of several crops, and this colonization can result in significant increase in the yield. Penyalver et al. [40] reported that *Agrobacterium rhizogen*s K84 is used worldwide as biocontrol agent against crown gall disease due to its multimechanisms of defense by the production of antibiotics, agrocin 84 and agrocin 434, and hydroxamate siderophores ALS84 as anti-agrobacterial substance. There is convincing evidence to support a direct role of siderophore-mediated iron competition in the biocontrol ability exhibited by bacterial isolates.The addition of a siderophore-producing *Pseudomonas putida* converted a *Fusarium*-conducive soil into a *Fusarium*-suppressive soil for the growth of three different plants. An isolate of *Pseudomonas cepacia*, positive for siderophore and β-1,3-glucanase production, decreased the incidence of diseases caused by *Rhizoctonia solani*, *Sclerotium rolfsii*, and *Pythium ultimum* [41].

In response to iron-deficiency stress, graminaceous plant species differ widely. Understanding the mechanism of stress responses is significant for increasing crop yields on calcareous soils. It also helps in improving the iron content of grains for human consumption. The response of graminaceous plants to iron deficiency occurs by the exudation of phytosiderophores to increase the availability of iron and by improving the uptake capacity of iron (III)-phytosiderophores. Phytosiderophores are usually hexadentate ligands that coordinate iron (III) with their amino and carboxyl groups. Phytosiderophores chelate sparingly soluble soil iron by forming iron (III)-phytosiderophore complexes that can be subsequently transported across the root plasma membrane via facilitated transport when released to the rhizosphere. In general, plant species releasing high quantities of phytosiderophores, such as barley, rye, and wheat are more resistant to iron deficiency chlorosis than species releasing smaller quantities, such as maize, sorghum, and rice. However, the quantity of phytosiderophores released is not always constant, for example, chlorosis resistance in different maize cultivars has been reported but this is not related to the total amounts of phytosiderophores released, indicating the contribution of other factors regulating iron efficiency process [42].

3. Identification of antagonistic antimicrobial producing rhizobacteria

Identification of bacteria is traditionally performed by isolation of the organisms and study of their phenotypic characteristics, including Gram staining, morphology, culture requirements,

http://dx.doi.org/10.5772/64869

and biochemical reactions. The discovery of PCR and DNA sequencing, comparison techniques of the gene sequences of bacterial species, proved that the 16S rRNA gene is highly conserved within a species and among species of the same genus, and thus can be used for bacterial identification at species level. For bacterial systematic studies at the family, genus, species, and subspecies levels, the 16S rDNA, which codes for the small subunit of ribosomal RNA, is now the most widely and successfully used informational macromolecule. For natural relationships between distantly related species and variable regions that can be used to separate closely related genera, the 16S rDNA conserved sequences can be used by constructing and comparing phylogenetic trees.

Such a 16S rDNA sequence-based identification technique will substantially facilitate the ecological study and the control of microorganisms difficult to culture [43].

Interests in biological control have recently increased due to imminent bans on chemical control, widespread development of fungicide resistance in pathogens, and a general need of sustainable disease control strategies. A wide variety of antagonistic biocontrol agents, such as *Pseudomonas, Burkholderia,* and *Trichoderma* spp., have been successfully identified, characterized, and utilized against many plant pathogens [44]. Now the agroindustry must focus on the identification and development of effective biocontrol agents against multiple pathogens as well as to develop the formulations that provide stable shelf-life and efficacy, and persistent user-friendliness.

Biocontrol of plant pathogens is being so popular because it can decrease the disease incidence, reduce the use of chemical fungicides, has no undesirable effects on nontarget organisms and environment, and is safer for the user and community.

4. Conclusions and scope

The plant growth promoting (IAA production, nitrogen fixation, and P-solubilization) and biocontrol traits (production of HCN, siderophores, hydrolytic enzymes, and antibiotics) suggest that these traits are more worthy of screening for plant growth promotion and bioantagonistic potential against plant pathogens. The plant growth-promoting rhizobacteria produce a wide variety of antimicrobial compounds against pathogens. A biocontrol agent possessing multimechanism systems of defense can antagonize root pathogens in a better way. This chapter highlights the need of screening the PGPR capable of producing a wide variety of antimicrobial compounds. Further evaluating/characterizing the biocontrol mechanisms and then testing the efficacy of selected antimicrobial-producing bacteria by lab, green house, and field trials could make them potent and successful biocontrol agents against many plant pathogens. This research chapter will help to minimize the chances of failure of biocontrol activity under field conditions, which is an emerging current problem of agriculture sector, and these tools will allow the isolation of improved antimicrobial bacterial strains and more efficient bioformulation to control pathogens. Molecular methods developed for the study of microorganisms in their environments are key tools for the study of the influence of the microbial community on biocontrol through variety of antimicrobial compounds produced by rhizobacteria. Further experiments should be initiated to study the optimum formulation and the interaction of these bacteria with the constituent of established PGPR preparations, with a view to incorporating them for field use. Research along these lines will increase the impact of PGPR on the biocontrol of plant diseases in the commercial world.

Acknowledgements

The author acknowledges Dr. Fauzia Yusuf Hafeez, COMSATS University Islamabad, Pakistan, for helping in the collection of the literature (books and journals) about antimicrobial rhizobacteria and antagonistic biocompounds and her expertise advise during the preparation of this chapter.

Author details

Munazza Gull

Address all correspondence to: munagull@hotmail.com

Biochemistry Department, Faculty of Science, King Abdulaziz University, Jeddah, Saudi Arabia

References

[1] Welbaum G, Sturz AV, Dong Z, Nowak J: Fertilizing soil microorganisms to improve productivity of agroecosystems. Critical Reviews in Plant Sciences. 2004; 23: 175–193.

[2] Flahaut S, Hartke A, Giard J, Benachour A, Boutibonnes P, Auffray Y: Relationship between stress response towards bile salts, acid and heat treatment in *Enterococcus faecalis*. FEMS Microbiology Letters. 1996; 138: 4944.

[3] Compant S, Duffy B, Nowak J, Clement C, Barka EA: Use of plant growth promoting bacteria for biocontrol of plant diseases: Principles, mechanisms of action and future prospects. Applied and Environmental Microbiology. 2005; 71: 4951–4959.

[4] Hallman J, Quadt-Hallman A, Mahafee WF, Kloepper JW: Bacterial endophytes in agricultural crops. Canadian Journal of Microbiology. 1997; 43: 895–914.

[5] Gull M, Hafeez FY, Saleem M, Malik KA: Phosphate-uptake and growth promotion of chickpea (*Cicer arietinum* L.) by co-inoculation of mineral phosphate solubilizing

bacteria and a mixed rhizobial culture. Australian Journal of Experimental Agriculture. 2004; 44: 623–628.

[6] Mayak S, Tirosh T, Glick BR: Plant growth-promoting bacteria confer resistance in tomato plants to salt stress. Plant Physiology and Biochemistry. 2004; 42: 565–572.

[7] Chatterton S, Sutton JC, Boland GJ: Timing *Pseudomonas chlororaphis* applications to control *Pythium aphanidermatum, Pythium dissotocum* and root rot in hydroponic peppers. Biological Control. 2004; 30: 360–373.

[8] Merharg AA, Killham K: Loss of exudates from the roots of perennial ryegrass inoculated with a range of microorganism. Plant and Soil. 1995; 170: 345–349.

[9] Bacilio-Jiménez M, Aguilar-Flores SE, Ventura-Zapata E, Pérez-Campos S, Bouquelet, Zenteno E: Chemical characterization of root exudates from rice (*Oryza sativa*) and their effects on the chemotactic response of endophytic bacteria. Plant and Soil. 2003; 249: 271–277.

[10] Dekkers LC, Mulders IH, Phoelich CC, Chin-A-Woeng TFC, Wijfjes AH, Lugtenberg BJJ: The *sss* colonization gene of the tomato-*Fusarium oxysporum* f. sp. *radicis-lycopersici* biocontrol strain *Pseudomonas fluorescens* WCS365 can improve root colonization of other wild-type *Pseudomonas* spp. bacteria. Molecular Plant-Microbe Interaction. 2000; 13: 1177–1183.

[11] Pierson LS, Pierson EA: Phenazine antibiotic production in *Pseudomonas aureofaciens*: role in rhizosphere ecology and pathogen suppression. FEMS Microbiology Letters. 1996; 136: 101–108.

[12] Haas D, Keel C, Reimmann C: Signal transduction in plant-beneficial rhizobacteria with biocontrol properties. Antonie van Leeuwenhoek. 2002; 81: 385–395.

[13] Kim BS, Moon SS, Hwang BK: Isolation, identification and antifungal activity of a macrolide antibiotic, oligomycin A produced by *Streptomyces libani.* Canadian Journal of Botany. 1999; 77: 850–858.

[14] Bloemberg GV, Lugtenberg BJJ: Molecular basis of plant growth promotion and biocontrol by rhizobacteria. Current Opinion in Plant Biology. 2001; 4: 343–350.

[15] Brodhagen M, Henkels MD, Loper JE: Positive autoregulation and signaling properties of pyoluteorin, an antibiotic produced by the biological control organism *Pseudomonas fluorescens* Pf-5. Applied and Environmental Microbiology. 2004; 70: 1758–1766.

[16] Ownley BH, Duffy BK, Weller DM: Identification and manipulation of soil properties to improve the biological control performance of phenazine-producing *Pseudomonas fluorescens.* Applied and Environmental Microbiology. 2003; 69: 3333–3343.

[17] Pettersson M, Baath E: Effects of the properties of the bacterial community on pH adaptation during recolonization of a humus soil. Soil Biology and. Biochemistry. 2004; 36: 1383–1388.

[18] Smith KP, Handelsman J, Goodman RM: Genetic basis in plants for interactions with disease suppressive bacteria. Agricultural Science. 1999; 96: 4786–4790.

[19] Lim HS, Kim YS, Kim SD: *Pseudomonas stutzeri* YPL-1 genetic transformation and antifungal mechanism against *Fusarium solani,* an agent of plant root rot. Applied and Environmental Microbiology. 1991; 57: 510–516.

[20] Kamensky M, Ovadis M, Chet I, Chernin L: Soil-borne strain IC14 of *Serratia plymuthica* with multiple mechanisms of antifungal activity provides biocontrol of *Botrytis cinerea* and *Sclerotinia sclerotiorum* diseases. Soil Biology and Biochemistry. 2003; 35: 323–331.

[21] Corbell N, Loper JEA: Global regulator of secondary metabolite production in *Pseudomonas fluorescens* Pf-5. Journal of Bacteriology. 1995; 177: 6230–6236.

[22] Zhang L, Birch RG: Biocontrol of sugar cane leaf scaled disease by an isolate of *Pantoea dispersa* which detoxifies albicidin phytotoxins. Letters in Applied Microbiology. 1996; 22: 132–136.

[23] Schouten A, van der Berg G, Edel-Hermann V, Steinberg C, Gautheron N, Alabouvette C, de Vos CH, Lemanceau P, Raaijmakers JM: Defense responses of *Fusarium oxysporum* to 2,4-diacetylphloroglucinol, a broad-spectrum antibiotic produced by *Pseudomonas fluorescens.* Molecular Plant-Microbe Interaction. 2004; 17: 1201–1211.

[24] Castillo UF, Strobel GA, Ford EJ, Hess WM, Porter H, Jensen JB, Albert H, Robison R, Condron MAM, Teplow DB, Steevens D, Yaver D: Munumbicins, wide-spectrum antiobiotics produced by *Streptomyces* NRRL 30562, endophytic on *Kennedia nigriscans.* Microbiology. 2002; 148: 2675–2685.

[25] Aino M, Maekawa Y, Mayama S, Kato H: Biocontrol of bacterial wilt of tomato by producing seedlings colonized with endophytic antagonistic pseudomonads. In: Ogoshi A, Kobayashi K, Homma Y, Kodama F, Kondo N, Akino S, editors. Plant Growth Promoting Rhizobacteria: Present Status and Future Prospects. Nakanishi Printing, Sapporo, Japan; 1997; p. 120–123.

[26] Ryasl JA: Systemic acquired resistance. Plant Cell. 1996; 8: 1809–1819.

[27] Penninckx IA, Eggermont K, Terras FR, Thomma BP, De Samblanx. GW, Buchala A, Metraux JP, Manners JM, Broekaert WF: Pathogen induced systemic activation of a plant defense gene in *Arabidopsis* follows a salicylic acid independent pathway. Plant Cell. 1996; 8: 2309–2323.

[28] Thomma BP, Eggermont K, Penninckx IA, Mauch-Mani B, Vogelsang R, Cammue BP, Broekaert WF: Separate jasmonate-dependent and salicylate-dependent defense-

response pathways in *Arabidopsis* are essential for resistance to distinct microbial pathogens. Proceedings of the National Academy of Sciences USA. 1998; 95: 15107–15111.

[29] Pieterse CMJ, van Wees SCM, van Pelt JA, Knoester M, Laan R, Gerrits H, Weisbeek PJ, van Loon LC: A novel signaling pathway controlling induced systemic resistance in *Arabidopsis.* Plant Cell. 1998; 10: 1571–1580.

[30] Compant S, Reiter B, Sessitsch A, Nowak J, Clément C, Barka EA: Endophytic colonization of *Vitis vinifera* L. by a plant growth-promoting bacterium, *Burkholderia* sp. strain PsJN. Applied and Environmental Microbiology. 2005; 71: 1685–1693.

[31] Gull M, Hafeez FY: Characterization of siderophore producing bacterial strain *Pseudomonas fluorescens* Mst 8.2 as plant growth promoting and biocontrol agent in wheat. African Journal of Microbiology Research. 2012; 6:6308–6318.

[32] Berg G, Roskot N, Steidle A, Eberl L, Zock A, Smalla K: Plant-dependent genotypic and phenotypic diversity of antagonistic rhizobacteria isolated from different *Verticillium* host plants. Applied and Environmental Microbiology. 2002; 68: 3328–3338.

[33] Ratledge C, Dover LG: Iron metabolism in pathogenic bacteria. Annual Reviews of Microbiology. 2000; 54: 881–941.

[34] Schwyn B, Neilands JB: Universal chemical assay for the detection and determination of siderophores. Analytical Biochemistry. 1987; 160: 47–56.

[35] Csaky TZ: On the estimation of bound hydroxylamine in biological materials. Acta Chemica Scandinavica. 1948; 2:450–454.

[36] Arnow LE: Colorimetric determination of the components of 3,4-hydroxyphenylalanine–tyrosine mixtures. Annual Review of Biochemistry. 1937; 50: 715–731.

[37] Witter AE, Luther WG: Variation in Fe-organic complexation with depth in the Northwestern Atlantic Ocean as determined using a kinetic approach. Marine Chemistry. 1998; 62: 241–258.

[38] Haselwandter K, Gunther W: Ferricrocin-an ectomycorrhizal siderophore of *Cenococcum geophilum*. BioMetals. 2002; 15: 73–77.

[39] Kloepper JW, Leong J, Teintze M, Schroth MN: Enhanced plant growth by siderophores produced by plant growth promoting rhizobacteria. Nature. 1980; 286: 885–886.

[40] Penyalver R, Oger RP, Lopez MM, Farrand SK: Iron binding compounds from *Agrobacterium* spp: Biological control strain *Agrobacterium rhizogens* K84 produces a hydroxamate siderophore. Applied and Environmental Microbiology. 2001; 67: 654–664.

[41] Frielender M, Inbar J, Chet I: Biological control of soil borne pathogens by a β-1,3-glucanase producing *Pseudomonas Cepacia*. Soil Biology and Biochemistry. 1993; 25: 1211–1221.

[42] Von Wirén N, Mori S, Marschner H, Römheld V: Iron inefficiency in maize mutant ys1 (*Zea mays* L. cv yellow-stripe) is caused by a defect in uptake of iron phytosiderophores. Plant Physiology. 1994; 106: 71–77.

[43] Olsen GJ, Woese CR: Ribosomal RNA: a key to phylogeny. FASEB Journal. 1993; 7:113–123.

[44] Kazempour MN: Biological control of *Rhizoctonia solani*, the causal agent of rice sheath blight by antagonistic bacteria in green house and field conditions. Plant Pathology Journal. 2004; 2: 88–96.

Chapter 3

The Organic Amendment Improve the Yield and Quality of Vegetable

Xingxue Huang, Aihua Wang, Bincai Wang, Guolin Zhou, Chufa Lin, Yaohua Deng and Runhua Zhang

Additional information is available at the end of the chapter

http://dx.doi.org/10.5772/64874

Abstract

Using biotechnology, we can change agricultural wastes into high-quality organic fertilizers, which leads us in the direction of the development in modern agriculture and act as substitute to the chemical fertilizers. In this chapter, five types of technologies of organic amendment are elaborated. Each method can be selected based on the specific circumstance. The effects of the technology in the production are introduced and the principles of the technologies are explained in a simple manner.

Keywords: organic amendment technology, vegetable yield, biological bacterium, organic fertilizers, vegetable quality

1. Introduction

1.1. The significance of organic amendment in vegetable production

The extensive use of fertilizers and pesticides leads to problems of environmental quality and product quality and safety. A decline in productivity, soil salinization, and groundwater pollution problems are paid more and more attention [1]. The primary care taken by customers is pesticide residue or food safety problems [2]. A large number of discarded agricultural wastes such as manure, crop straw, and product residues lead to a lot of spoilage of microorganisms and bacteria. Burned crop straw pollutes the air. Using biotechnology, we can change agricultural wastes into high-quality organic fertilizer, which leads us in the direction of development in modern agriculture.

2. The ways of organic amendment

2.1. Returning straw into field

Due to the development of mechanization, returning straw into field (straw returning) is promoted throughout the world. Straw returning technology has been accepted by farmers. Straw crushing, *in situ* and returning to field, can not only improve the soil fertility and soil structure, increase soil organic matter, and income, but also reduce air pollution caused by burning straw.

Annual production of agricultural straw in China is more than 600 million ton, which is a large amount of wealth. In recent years, with the development of high horsepower tractors and harvesting machinery, the straw crushing and deep plowing and returning are developing. Straw *in situ* crushing and returning to field is more convenient and labor saving, and popular. However, the technology must be used properly. The corresponding technical measures and fermentation strains must be set [3].

Some understanding and experience for straw return to field are explained.

2.1.1. The time when corn straw returning to the field should be as early as possible

The moisture content of green corn straw is higher than the dry straw, so the straw crushed and buried into soil is easier to decompose. The best time to return corn straw to the field is just after harvest.

2.1.2. The appropriate amount of straw returned

Although the straw returned into field can improve fertility, the amount of straw returned to the field is important, the lesser, the better. The proper standard is to cover the ground with straw if there are no appropriate measures accelerating the fermentation. In general, for maize straw one half of an acre of straw is returning amount for an acre of soil. When the ground is not covered entirely by the straw, the effects of water conservation and weed suppression are not obvious, whereas with excessive amount, the straw does not decompose and there will be some difficulties in deep plowing.

2.1.3. Crushing length of straw returned

Because the withered corn subtending leaf is not easy to be crushed, the leaf must be taken out of the field. Poor crushing effect means poor decomposition. To ensure the crushing effect, a tractor with large horsepower matching with straw returning machine should be chosen. Straw length of less than 4 cm is appropriate. When the straw is tilled deeply, the soil can be mixed evenly. The depth of plowing must not be less than 23 cm.

2.1.4. A sufficient amount of base fertilizer

Straw composting process will consume nitrogen and other available nutrients in the soil, it is appropriate to use some nitrogen fertilizer that decreases the straw carbon and nitrogen ratio. The right carbon and nitrogen ratio is conducive to microbial activity and organic matter decomposition, so that the contradiction between microbes and crop competing nitrogen are addressed. Therefore, when the straw is returned into the field, it is necessary to keep sufficient nutrients in the soil, such as ammonium bicarbonate and calcium superphosphate fertilizer. If conditions permit, planter can substitute human excrement or poultry manure fertilizer for chemical fertilizer.

2.1.5. Application of fermentation bacteria

The fermentation bacteria include CM and BM bacteria. CM bacteria, from Biological Technology Co., Ltd. of Shanxi Yuncheng, are liquid biological bacteria, and BM bacteria are from Hebi Biological Technology Co., Ltd. The dosage is 5 kg CM bacteria and 5 kg BM bacteria for 5 tons of organic straw. Twenty-four hours before the application of biological agents, intermediate culture must be taken. Water, brown sugar, and biological agents (CM:BM = 1:1) with a ratio of 100:5:1 should be mixed. The sugar must be dissolved before the static culture for 24 hours. If necessary, the culture can be further expanded using brown sugar and water. The biological agents with intermediate culture, wheat bran, and water must be mixed with a ratio of 1:5:9, covering them with film for 5–7 hours, and should be evenly spread in the crushed straw.

2.1.6. Appropriate amount of lime

Because various organic acids produced during the composting process of fresh straw injure crop root, appropriate amount of lime in the vent and acidic soil should be applied, preferably in 400 kg/ha.

2.1.7. The straw with diseases and insect pest

In the plots with serious plant diseases and insect pests, the straw will not be suitable for direct returning because of infectious diseases, turning a large number of eggs of the pest and pathogen into the soil. Such straw can be used in composting with high temperature to kill bacteria, and also be used as feed and fuel.

2.2. Fermentation bioorganic fertilizers (organic fertilizer formula, technology application)

Poultry manure and crop straw are two kinds of organic wastes in China. China's annual livestock and poultry manure is about two billion tons, far more than the industrial solid waste. The chemical oxygen demand (COD) of poultry manure is far greater than that of the industrial wastewater and domestic wastewater. It is one of the main causes of pollution in rural areas. As a result of the disappearance of firewood, a large number of crop straws are burned, which cause air pollution and at the same time increase carbon emissions, resulting in the waste of large amount of organic carbon resource. On the other hand, the long-term application of chemical fertilizer has a bad influence on the soil ecology and the quality of the products. Good organic fertilizer is the urgent need of the hour.

In the existing technologies for fermentation of poultry manure and straw for production of organic fertilizer, there are the following shortcomings: complete fermentation needs a long time at low temperatures, which in return pollutes the air; manurial efficiency is low and cannot fully meet the needs of crop growth, still a quantitative amount of chemical fertilizer needs to be added.

2.2.1. *Preparation materials*

The fermentation bacteria include CM bacteria and BM bacteria. CM bacteria are liquid biological bacteria and are from Biological Technology Co., Ltd. of Shanxi Yuncheng, whereas BM bacteria are from Hebi Biological Technology Co., Ltd. The dosage is 5 kg CM bacteria and 5 kg BM bacteria for 10 tons organic fertilizers. Twenty-four hours before the application of biological agents, intermediate culture must be taken. Water, brown sugar, and biological agents (CM:BM = 1:1) must be mixed with a ratio of 100:5:1. The sugar is dissolved before the static culture for 24 hours. If necessary, the culture can be further expanded using brown sugar and water. The biological agents with intermediate culture, wheat bran, and water must be mixed with a ratio of 1:5:9, covering them with film for 5–7 hours, and evenly spread in the crushed straw [4].

The cow dung is fresh, whose water content is 50–60%.

The asparagus bean, sesame, and corn stalk are crushed into 1–3 cm filament.

Wheat bran is from the ordinary market sales.

2.2.2. *The method of making the biological organic fertilizer*

The cow dung and cowpea, sesame, and corn straw mentioned above are mixed by pile turning, and then spread out layer by layer, each layer is about 10–15 cm high; a layer of the wheat bran mixed with biological agents is sprinkled on each layer of the mixture. The layer of wheat bran is 0.5–1 cm thick. The completed compost is about 0.8–1.2 m high, and to facilitate the operation it is covered tightly with membrane for anaerobic fermentation for 3–5 days.

Three to five days after the fermentation, the compost is turned and piled up about 1.3–1.7 m high and 1.8–2.2 m wide. The compost is not covered, but it is protected from the rain. A spade handle ramp is used into the compost to form 2–3 vents per square meters. Note that 10–15 days after composting, fermentation is completed when white mycelium appears in the compost.

The compost could be used directly. Granular products can also be formed through milling and granulating, in which the final moisture content is 20–30%.

The compost should be covered with plastic membrane and sheltered from rain, so that the organic fertilizer can be saved for more than 2–3 years.

Compared with the prior art, the invention has the following advantages and effects:

1. Formula is reasonable, there are a variety of biological bacteria, and the fermentation can be completed at low temperature within 10–15 days.

2. The compost can improve the soil temperature and the CO_2 concentration in the greenhouse, thus improving the seedling growth at low temperature.

3. With high manorial efficiency, the organic fertilizer can fully meet the normal growth of plant, without adding chemical fertilizer.

4. The compost containing a variety of small molecular nutrients can maintain normal growth of crops in the weather with low temperature and weak light.

5. The compost or organic fertilizer can improve the yield by 20–120% and have earlier listing by 5–7 days compared with the chemical fertilizer under the conditions of greenhouse in early spring.

3. Straw reactor technologies

3.1. Six major roles of straw biological reactor

3.1.1. Effects of carbon dioxide

The reactor can generally increase carbon dioxide concentration by 4–6 times in plastic greenhouse, improve photosynthetic efficiency by more than 50%, accelerate growth, and improve rates of flowering and fruit setting. Standardized operation increases yields of cucumber and tomato by 30–80% [5].

3.1.2. Heat effect

In the greenhouse in cold winter, the temperature increases by 4–6°C 20 cm underground and air temperature increases by 2–3°C, thus improving plant growth environments and the ability of the crop to resist low temperature, effectively protecting the normal growth of crops and advancing the growth period by 10–15 days.

3.1.3. Biological control effect

Strains produced a large number of resistant spores in the conversion process of straw producing strong antagonistic, suppression, and lethal effect on plant diseases and insect pests. Plant disease rate is reduced by more than 90%, and the dosage of pesticide reduced by more than 90%. The standardized operation can be basically without pesticides.

3.1.4. Modifying the soil

In straw bioreactor planting layer of 20 cm underground, soil porosity increased 1 times, and also the beneficial microbial groups. The conditions of water, fertilizer, gas, and heat is medium, various mineral elements are directionally released, and organic matter content increased more than 10 times, which creates a good environment for root growth.

3.1.5. Decreasing pesticide residues

In the reaction process, the flora metabolism produced a large number of high activity of the enzyme, which reacted with chemical fertilizer and pesticide, so invalid fertilizer becomes effective, the harmful substances becomes beneficial, and eventually make pesticide residue into carbon dioxide which plants need. It was determined that the pesticide residues in the soil around the plant roots were decreased by more than 95% in 1 year, and eliminated in 2 years.

3.1.6. Improving the comprehensive utilization of natural resources

Straw bioreactor technology speeds up the use of straw, while improving the comprehensive utilization of natural resources such as the microbial, light, water, air, and other natural resources. According to the measurement, the carbon dioxide concentration increased four times, the light utilization rate increased by 2.5 times, water use efficiency increased by 3.3 times, and legume nitrogen fixation activity increased by 1.9 times.

3.2. Application methods and key points of four straw bioreactor technology

There are three main ways to operate the technology: internal, external, and internal-external bioreactors. Selection of application methods mainly depended on the production of crop varieties, planting time, ecological climate characteristics, and production conditions.

The choice and condition of the internal straw biological reactor:

1. For internal straw biological reactor under row: in autumn, winter, and spring season it can be used, and in high altitude, high latitude, drought, cold, and short frost-free areas it should be used, especially.

2. The internal straw biological reactor between rows: in high temperature season and in the area where there are no straw it should be used before planting.

3. The internal type for topdressing: the whole process of crop growth can be used, and the method is more flexible. Straw should be crushed and applied in holes.

4. The internal type under tree: in fruit trees, economic forest, green belt, and nursery planting areas it should be adopted

The straw, stain, and excipients of the internal straw biological reactor:

1. For internal straw biological reactor under row: every 667 m^2 the amount of straw 3000–4000 kg, strain 8–10 kg, wheat bran 160–200 kg, cake fertilizer 80–100 kg

2. The internal straw biological reactor between rows: every 667 $m^2$the amount of straw to be used is 2500–3000 kg, strain 7–8 kg, wheat bran 140–160 kg, and cake fertilizer 70–80 kg.

3. The internal straw biological reactor for topdressing: every 667 m^2 each straw powder (or edible fungus waste) dosage is 900–1200 kg, strain 3–4 kg, wheat bran 60–80 kg, and cake fertilizer 80–100 kg.

4. The internal straw biological reactor under the tree: every 667 m^2 the amount of straw to be used is 2000–3000 kg, strain 4–6 kg, wheat bran 80–120 kg, and cake fertilizer 60–90 kg.

3.2.1. *The processing method of strain*

Prior to the day when using or on the day, the strain must be pretreated. Methods: 1 kg strain blending 20 kg of wheat bran, 10 kg of cake, adding 35–40 kg of water, and after 4–24 hours of fermentation it can be used. If it is not over, the mixture should be spread in the room or shade, keeping a thickness of about 8–10 cm, and continue to use on the next day.

3.2.2. *Attention*

Razing animals' (cattle, horses, sheep, etc.) feces can be used for growing vegetables, fruits, and legumes: in the use of fertilizer technology grazing animal manure should not be used. Research confirmed that the use of chicken, pig, human, ducks, and other nonherbivoros animal manure will accelerate the nematode reproduction and dissemination, causing plant diseases; use of fertilizers will influence the activity of bacteria, also can make soil compaction and accelerate the disease spread.

3.2.3. *The operation of the internal straw biological reactors*

Ditching and laying straw, sprinkling strain, vibration, covering soil, watering, plotting, drilling, and planting.

1. Ditch: double line should be used for planting size. Big lines (sidewalk) are 100–120 cm wide, little lines are 60–80 cm wide; ditching furrow under the little lines. They are 60 or 80 cm wide, 20–25 cm deep, as long as the planting line. The soli dig up is put on each side of the furrow.

2. After ditching, paving straw in the trench (corn straw, wheat straw, and rice straw, etc.). At the bottom of the general shop put the whole straw (corn straw, sorghum straw, firewood, etc.), at the top with broken soft straw (e.g., rice straw, wheat straw, corn bran, weeds, leaves, and edible mushroom leftover). The straw put down and compacted is 25–30 cm thick, at both ends of the ditch straw stubble exposed 10 cm in order to enter the oxygen.

3. Sprinkling strains: in each furrow 6 kg of treated strains are evenly sprinkled on the straw, and patted again by a shovel, so the strains and straw are closely contacted.

4. Covering soil: putting the soil on both sides of ditch back into ditch. Soil is 20–25 cm thick, planting ridge is formed, and the ridge is leveled.

5. Watering: watering to split the straw, 3–4 days later, leveling the ridge surface and the soil is kept about 20 cm thick.

6. Drilling: on the ridges with reinforcement 12# (generally 80–100 cm), and in the top welding a T type to drilling three rows of holes, line spacing 25–30 cm, and hole spacing 20 cm, to penetrate the straw layer, letting oxygen in, and promoting the straw fermentation, standing by for planting.

7. Planting: generally pouring water, a bowl of a tree. Watering 3 days after planting under high temperature, and 5–6 days under low temperature.

3.2.4. The internal straw biological reactor between rows:

1. Ditch: double line should be used for planting size. Big lines (sidewalk) are 100–120 cm wide, little lines are 60–80 cm wide; ditching furrow under the little lines. They are 60 or 80 cm wide, 15–20 cm deep, as long as the planting line. The soli dig up is put on each side of the furrow.

2. After ditching, paving straw in the trench (corn straw, wheat straw, and rice straw, etc.). At the bottom of the furrow put the whole straw (corn straw, sorghum straw, firewood, etc.), at the top with broken soft straw (e.g., rice straw, wheat straw, corn bran, weeds, leaves, and edible mushroom leftover). The straw put down and compacted is 20–25 cm thick, at both ends of the ditch straw stubble exposed 10 cm in order to enter the oxygen.

3. Sprinkling strains: in each furrow 6 kg of treated strains are evenly sprinkled on the straw, and patted again by a shovel, so the strains and straw are closely contacted.

4. Covering soil: putting the soil on both sides of ditch back into ditch. Soil is 20–25 cm thick, planting ridge is formed, and the ridge is leveled.

5. Watering: watering the little line (plant line) to percolate into big line (sidewalk), 3–4 days later, leveling the ridge surface and the soil is kept about 20 cm thick.

6. Drilling: on the ridges with reinforcement 12# (generally 80–100 cm), and in the top welding a T type to drilling three rows of holes, line spacing 25–30 cm, and hole spacing 20 cm, to penetrate the straw layer, letting oxygen in, and promoting the straw fermentation, standing by for planting.

7. Planting: generally pouring water, a bowl of a tree. Watering 3 days after planting under high temperature, and 5–6 days under low temperature.

3.2.5. The internal straw biological reactor for topdressing

In order to maintain production of the whole growing period, the method should be used in the growth period. The new straw is crushed, adding the mixture every 667 square including 3 kg bacteria, wheat bran 60 kg, cake 30 kg, straw powder 900 kg, and water 2000 kg (the proportion of 1:20:10:300:666). The trapezoidal reactor mixture is piled up as high as 60 cm, and as wide as 100 cm to ferment. By the sticks with a diameter of 5 cm drilled the nine holes on the pile surface. Covering with membrane fermentation, when the compost is heated to 45°C to 50°C, it can be used to fertilize the caves dug in the soil. The caves are 15 cm from the plant and 30 cm from each other. A total of 0.5–1.0 kg compost per cave is fertilized; after

http://dx.doi.org/10.5772/64874

covering the soil, 3–4 per holes are drilled on the cave; no watering is done for 7–10 days after topdressing, while depending on the soil moisture content during the growing period watering is done 2–3 times.

3.2.6. *The internal straw biological reactor for trees*

According to the different application period it is divided into full and half biological reactor, it is suitable for fruits, also green trees, antisarin, and other species of higher value may be used as reference.

The full one: The furrow is around the trunk from the surrounding soil to crown projection below, which is 10–25 cm deep. Most capillary roots may be exposed or broken. The furrow is sprinkled by a layer of vaccine, and covered with straw that is 10 cm higher than the ground, 10 cm stalks are exposed out of the pit for oxygen. The soil is filled back. Irrigate enough water, level the soil, punch, and covered film 3–4 days later. With reinforced 12#, the holes are punched with 30 × 25 cm after germination.

The half one: The method is applied in the growing season of fruit trees. Practice is to be around the trunk at six equal parts, the furrow is fan shaped and 40–60 cm deep (preventing root injury). Sprinkle a layer of vaccine, and then lay the half straw, sprinkle a layer of bacteria, lay the other straw, sprinkle another layer of bacteria, pat the soil with a shovel, and 3 days before watering and leveling the soil, punch the hole to 30 × 30 cm^2. It does not cover the plastic film, but cover plastic film in the plateau area with water shortage to protect water. Its operation method is as same as the internal straw biological reactor.

3.2.7. *The external application of straw bioreactor*

According to the level of investment and construction quality it can be classified into simple external and external standard. Simple external type: it only need to dig trenches, lay thick plastic sheeting on it, do isolation layer with sticks, small cement pole, bamboo billet or branches, build the base with brick, cement for airway, and switch base. It is characterized by small investment, fast construction, but the film is easily damaged.

Standard external: trenching, construct gas storage pool, airway, and the switch base with cement, brick, and sand, do isolation layer with cement poles, bamboo billet, and gauze. Although, the investment is large, the period of use is long. According to its construction site, in the low temperature season, it is built in the shed, and outside the shed in high temperature season. The one outside is convenient. The construction process should be built before sowing or planting. The feeding should be after planting or seedling.

The straw, strains, and auxiliary materials dosage: each time: straw, 1000–1500 kg, strain, 3–4 kg, wheat bran, 60–80 kg. The whole growth period:2–4 times.

Construction period: during the whole growth period the application of external type biological reactor increases yield, the sooner the reactor is used, the larger the yield is increased. Average increase of yield is more than 50%.

The construction process of external reactor:

1. Standard external: for winter and early spring crop it is built in at the inner side of the gable greenhouse imported, 60–80 cm away from the gable, from north to south to dig 120–130 cm for a catchy width and 100 cm deep, 90–100 cm for mouth wide, long, 6–7 m (a little less than greenhouse width) of ditch. The excavated soil is evenly placed along the ditch, and the shape of the outer high and low profile is spread out. The ditch is laid with plastic sheeting (can reduce the dosage of cement and sand walls), and extends 80–100 cm along the trench. And then from the middle of the ditch excavate an airway that is 65 cm wide, 50 cm deep, 100 cm long, connected with the ground round the exchange base. The base diameter is 50 cm. At the two ends of the ditch, the back airway is constructed with a length of 50 cm, width of 20 cm, and height of 20 m. Then feeding inoculation begins: sprinkle three layers of straw and bacteria (every layer: 40–50 cm thick straw, with a layer of bacteria), wet straw so that half of furrow is filled with water. Finally, the furrow is covered with plastic sheeting to cover the moisture. The covering should not be too strict, the same day pumping gas, so that gas can be circulated to accelerate the reaction.

2. The simple external construction process: ditching with external standard. Just in order to save ditch costs, cement, sand, and brick of the ditch wall is replaced with plastic sheeting.

3.2.8. The use and management of external reactor

External reactor use and management can be summarized as: "three application" and "three amendments." Boot uninterruptedly on the day of feeding regardless of that the weather is cloudy or sunny.

Gas: to boot for 5–6 hours per day at the seedling stage, 7–8 hours at the flowering period, and 10 hours per day at the fruit period, whether it is cloudy or sunny. It is confirmed that the reactor carbon dioxide gas can increase production by 55–60%, especially at noon.

Liquid: on the second day water in the ditch should be irrigated out of the ditch, poured the leaching in the straw three times on the reactor, once a day. If the ditch is short in water, replenish it. The reason is that the activity of enzyme and spore is high, and the effect is good. The solution including 1 parts of the leaching solution and 2–3 parts of water is sprayed on the leaves or roots, 3–4 times per month. Reactor leaching solution contains a large amount of carbon dioxide, mineral elements, and disease-resistant spores, which can increase plant nutrition, and it can also play the role in prevention and control of diseases and insect pests. It is proved that the reactor liquid can increase the yield by 20–25%.

Slag: straw reactor releases a large number of mineral elements, except dissolved in leaching solution and also in slag. It is a mixture of organic and inorganic nutrients vegetables need. External reactor slag can be cleaned out and piled up to decay into powder as matrix fertilizing, it is to not only to replace chemical fertilizers for seedling growth, but also to prevent and control plant diseases and eliminate damage from pests. It is showed that the reactor slag can increase the yield by 15–20%.

The water: water is one of the important conditions of the reactor. In addition of water in building a pile, the water must be supplied every 7–8 days to fill a water reactor. Untimely replenishment will reduce the efficiency of the reactor, causing the reactor to stop.

The oxygen is a prerequisite for the reactor to produce carbon dioxide. The straw bioreactor needs a large number of oxygen. With the reaction reactor working, ventilation condition is getting worse and reaction is slow so that the covering film on the piles should not be too strict, every 20 days the film should be uncovered, and 5–6 hole per square meter should be drilled.

Amendments: when external reactors generally use for 50 days or so, the straw consumption is more than 60%. Note that 1200–1500 kg straw and strains should be added. In winter the amendments are supplied three times.

3.2.9. *The notice for operation*

1. The operating time of the internal reactor should be more than 20 days ahead of planting, otherwise the results will be postponed.
2. The first watering should be enough to foot (with wet straw); the second watering is uniform, with the interval time of 10–15 days; watering for the third time to be clever, watering for the fourth time should be careful, in winter or spring period it should not be watered, if not be dry.

The use of built-in master four should not be the principle of:

Ditching should not be too deep (not to exceed 25 cm); strain and straw quantity should not be too little (straw 3000–4000 kg, 8–10 kg bacteria per 667 m^2), covering soil should not be too thick (20–25 cm); and drilling the hole should not be too late, or too little (3 days after watering, a hole per 20 m^2).

3.3. The application effects of straw biological reactor technology

3.3.1. *Growth performance*

Seedling: early onset—fast growth, wide stem diameter, short internodes, large and thick leaves, early flowering, fewer pests and diseases, and resisting natural disasters. Medium—strong growing of crop, many big fruit of less deformity, and 10–15 days of listing period in advance. Late—the strong ability of continuous fruiting, longer harvest period by 30–45 days, and obstacles lead to continuous cropping are overcome.

3.3.2. *Yield performance*

The yield of different fruit varieties generally increase by 80–500%; the yield of different vegetable varieties increase by more than 50–200%, the yield of root, stem, and leaf crops generally increase 1–3 times, the yield of legumes (such as peanuts and soybeans) increase by 50–150%.

3.3.3. *Quality performance*

Fruit tidy, commodity rate, the color and luster, sugar content, flavor, and aroma quality are improved significantly; nitrite content and pesticide residues significantly decreased or disappeared.

3.3.4. *The input-output ratio*

Greenhouse vegetable, melons for 1:14–16; high arch shed vegetables, melons for 1:8–12; small arch shed melons, vegetables 1:5–8; open field cultivation of melon and vegetable 1:4–5; and special Chinese village is 1:20–50.

3.3.5. *Reducing production costs*

In greenhouse the reduction is 3500–4000 yuan per 667 m^2; in shed it is 1500–2500 yuan; in small arch shed it is 500–1000 yuan.

4. Plant vaccines and their use

Plant vaccine is a kind of biological technology that uses plant immune function to prevent plant diseases. The mechanism of preventing and controlling the disease is to activate the immune function and to realize the purpose of preventing and controlling the disease. It is an important part of the biological reactor technology system. The technology is now used in Shandong, Liaoning, Hebei, and other 10 provinces, as well as in more than 100 counties (cities, districts); fruit trees, vegetables, herbs, legumes, tea, tobacco, and other crops on large areas; also, the control effect reached 80–100%, the average cost decreased by 60%, with an average increase of more than 30%. The plant vaccine has important significance to solve the problem of pesticide pollution and pesticide residues in agricultural products, and to realize the organic cultivation and food safety of crops.

5. Production of nursery soil

5.1. How to make breeding soil with biological organic fertilizer

There are two stages in field cultivation of crops. In seedling stage it is mainly for crop root growth and seedling cultivation so that fertilizer quantity is not big, but uniform. There cannot be too high nutrients, otherwise it will produce burning phenomenon of seedling roots. The nutrients needed for the growth of seedlings were mainly supplied by the bed soil and the substrate. The bed soil or other nursery matrix requires loose, fertile, air permeability, and moisture retaining property.

Production of nursery soil: 10% organic biological fertilizer and 90% ripening fertile garden soil mixing, sieving, and joining right amount of nitrogen, phosphorus, and potassium

http://dx.doi.org/10.5772/64874

nutrient. Available nutrient contents are controlled as nitrogen 150–300 mg/kg, 200–500 mg P_2O_5/kg, and K_2O–600 mg/kg. Nursery soil adding fertilizer can be calculated according to the effective nutrient content, generally as for 100 kg nursery soil, 0.5 kg ammonium sulfate, 1 kg superphosphate, and 1 kg potassium sulfate, which are mixed evenly to avoid inhibition of seedling growth.

Acknowledgements

This research was supported by the Project of the National Modern Agriculture Industry Technology System (CARS-25-G-30). The authors thank the agricultural expert Song Zhixiong and Ma Xinli for the suggestions and confirmation on technology details and application effects in this manuscript.

Author details

Xingxue Huang*, Aihua Wang, Bincai Wang, Guolin Zhou, Chufa Lin, Yaohua Deng and Runhua Zhang

*Address all correspondence to: hxx329517@126.com

Wuhan Institute of Vegetable Science, Wuhan, China

References

[1] Xia Q. The effects of chemical fertilizers on the soil and its countermeasure. Agricultural Science of Guangdong, 2010, 09, 202–204.

[2] Mondelaers K, Verbeke W, Van Huylenbroeck G. Importance of health and environment as quality traits in the buying decision of organic products. British Food Journal. 2009, 111 (10), 1120–1139.

[3] Wang X, Youxiang L, Yan L, Lihua X. The progress of straw-returning technology. Modern Agriculture, 2015, 2, 1–4.

[4] Huang X, Lin C, Wang B, Hu Z, Zhou G, Aihua W, Yu C. Biofertilization alleviates problems of continuous cowpea cropping by degrading autotoxins in soil. Allelopathy Journal, 2014, 34 (2), 265–276.

[5] Ma X, editor. The talks of Ma Xinli on the highly effective cultivation of organic vegetable. 1st edition. Beijing: Science and Technology Documentation Press, 2009, pp. 69–127.

Chapter 4

Plant Pathogens

Waleed M. Abdulkhair and Mousa A. Alghuthaymi

Additional information is available at the end of the chapter

http://dx.doi.org/10.5772/65325

Abstract

Plants cover the most area of the earth's living environment as trees, grasses, flowers, and so on. Plants play different important roles in the environment such as ecosystem balance and food supplement for animals and humans. Moreover, wild or cultivated plants are considered the powerful biofertilizers for the soil, where the plant debris after death and degradation provides the soil with sufficient organic matters. Accordingly, plant care is a great duty and hard mission, which must be constantly improved. The study of plant pathogens belongs to the branch of biology known as plant pathology. The latter is also concerned to overcome the plant diseases arising from the biotic and/or abiotic origin. Biotic (infectious) diseases are developed owing to microbial infection, while abiotic (noninfectious) diseases are developed due to environmental factors. In this chapter, we are concerned with plant pathogens or phytopathogenic microbes such as bacteria, viruses, fungi, mollicutes, and so on.

Keywords: phytopathogens, biotic diseases, abiotic diseases, parasitism, pathogenicity

1. Introduction

The plant pathogens especially microbes will be the main subject of this chapter. The science, which is concerned with the study of plant diseases and their causes, is known as plant pathology. Therefore, all scientists concerned with this science constantly attempt to treat the diseased plants via various methods. This approach of scientific research is very important owing to the economic and hygienic yield for humans and animals. The phytopathogens are two types: biotic factors, which include all microbes and parasitic plants, and abiotic factors, which include all environmental factors. Essentially, the plant pathology is correlated with other sciences such as entomology, bacteriology, mycology, virology, and weed science due to deleterious effects of insects, bacteria, fungi, viruses/viroids, and weeds on plants, respectively. The first step of plant disease treatment is observation of definite and clear symptoms on the plants. These symptoms give an initial indication for the type and cause of plant disease, which may end with the death [1]. The modern approach of plant disease control depends on biological control agents such as the production of antimicrobial agents and the production of genetic-improved strains of plants, which are more resistant to plant diseases. This approach is more favorable because it is friendlier with the environment and healthier for humans and animals [2]. The infected part of the plant gives an indication of the type of plant disease, such as infected root which is usually correlated with root-rot disease [3]. The plant diseases can be classified according to several parameters: disease symptoms, infected organ, infected plant type, and the type of phytopathogen. The latter is considered the more useful criterion used for plant disease classification, because it easily determines the disease cause, potential disease complications, and possible control methods [4]. According to this criterion, plant diseases are classified into two types: infectious (biotic) diseases, which are caused by eukaryotes, prokaryotes, parasitic higher plants, viruses/viroids, nematodes, and protozoa, and noninfectious (abiotic) diseases, which are caused by different extreme environmental conditions [5].

2. Basic procedures in the diagnosis of plant diseases

The plant disease diagnosis depends on the exact determination of the disease cause. Generally, there are two plant disease causes: the pathogens and/or environmental factors. The former leads to infectious diseases, while the latter leads to noninfectious diseases [6].

2.1. Infectious diseases

There are wide range of phytopathogens which cause infectious plant diseases such as fungi, bacteria, viruses, viroids, mollicutes, parasitic higher plants, and protozoa. The infectious disease means the ability of phytopathogen to transfer from the infected plant to another healthy one and causes the same disease and the same symptoms. The most phytopathogens can inhabit the internal environment of plants; however, some others can live on the plant surface such as some fungi, bacteria, and parasitic higher plants [7].

2.1.1. Diseases caused by parasitic higher plants

Some plant diseases are developed due to growing certain plants attached on or in other plants, where they take all required nutrients without benefit sharing; these plants are called parasitic higher plants. This abnormal relationship leads to weakness of healthy or host plant. The parasitic higher plants are usually found attached with the surface of the host plant, such as dodder, mistletoe, witchweed, and broomrape [8].

2.1.2. Diseases caused by nematodes

The nematodes are one of most common phytopathogens which have definite symptoms. These symptoms only appeared in the infected site. The nematode infections in or on plants

http://dx.doi.org/10.5772/65325

are widely distributed especially in proper environments such as moderate temperature and high humidity [9].

2.1.3. Diseases caused by fungi

Interestingly, there are two main types of fungi appearing on plants: pathogenic and saprophytic. The pathogenic fungi live in or on plant tissues and cause serious complications for the vital physiological functions of plants, while saprophytic ones live in or on dead tissues. Accordingly, the diagnosis of plant disease must be exactly carried out. The exact diagnosis and determination of fungi take place by microscopical examination to identify the mycelial morphological characteristics, whatever fruiting structures and spores. After complete identification for the fungus and the symptoms of plant disease, the latter should be compared with that reported in the reference. This study will exactly determine whether the fungus is a pathogen or a saprophyte. Although microscopical examination is an essential and effective method for fungal identification, it only sometimes cannot lead to exact identification due to the absence of fungal fruiting structures and spores on infected plant tissue. Therefore, an alternative method must be used, such as using selective media for isolation, identification, or promotion of sporulation. On the other hand, some fungi need to be incubated under certain temperature, aeration, or light conditions to produce spores [10].

2.1.4. Diseases caused by bacteria and mollicutes

The appearance of bacterial growth in or on plant tissues means that bacterial plant disease may be present, because saprophytes may be present. Therefore, accurate bacterial identification must be carried out by using microscopical examination and physiological parameter determination. The selective media are essentially used in the bacterial identification to determine the bacterial genus and species in some cases. Moreover, the confirmatory test of bacterial pathogenicity may be carried out by inoculation of single pure bacterial colony in the healthy plant, reproducing the same symptoms that appeared on the infected one. Moreover, immunodiagnostic techniques or serodiagnostic assays can be used, such as agglutination and precipitation, fluorescent antibody staining, and enzyme-linked immunosorbent assay (ELISA). There are several advantages for these techniques such as quite sensitivity, fairly specificity, rapid, easy to perform, and it is expected that standardized, reliable antisera will be available soon. Furthermore, there are recent methods used for bacterial identification, which depend on the automated analysis of bacterial fatty acid profile. The molecular biological techniques are also widely used [11]. There are uncommon microorganisms called mollicutes. These microorganisms are very small where they must be examined by an electron microscope. Mollicutes have polymorphism and lack cell wall–like mycoplasma. These microorganisms habit the young phloem cells as a convenient host, and cause severe plant diseases such as plant stunting, yellowing or reddening of leaves, proliferation of shoots and roots, production of abnormal flowers, and eventual decline and death of the plant. Mollicutes cannot be cultured on nutrient media except for the genus *Spiroplasma*. Mollicutes can be diagnosed by several parameters, such as symptoms determination, grafting, transformation, microscopical examination, susceptibility to tetracyclines, and so on [12].

2.1.5. Diseases caused by viruses and viroids

There are distinctive types of plant diseases caused by viruses/viroids. These diseases have definite and clear symptoms, which easily support disease diagnosis and are considered main advantage. Apart from this advantage, some recent techniques are widely used for disease diagnosis and virus identification, such as virus transmission tests to specific host plants by sap inoculation, grafting, certain insect, nematode, fungus, and mite vectors. Moreover, serodiagnostic tests are used for this purpose such as enzyme-linked immunosorbent assays, gel diffusion tests, micro-precipitin tests, and fluorescent antibody staining. The electron microscopy techniques as negative staining of virus particles in leaf dip or purified preparations are also used, as well as immune-specific electron microscopy. On the other hand, there are more accurate techniques used for disease diagnosis and virus/viroid identification, such as electrophoretic tests and hybridization of commercially available radioactive DNA complementary to a certain virus DNA or RNA, or viroid RNA, with the DNA or RNA present in plant sap and attached to a membrane filter (immunoblot) [13].

2.1.6. Diseases caused by more than one pathogen

Sometimes, some plants are exposed to coinfection by two or more pathogens, which lead to the same or different disease symptoms. Therefore, the differentiation and identification of these pathogens are very essential to exactly determine the disease cause. The differentiation and subsequently identification are carried out by all techniques that are mentioned above [14].

3. Noninfectious diseases

Occasionally, some plant diseases have abiotic origin such as environmental factors; these diseases are called noninfectious diseases. Abiotic environmental factors have deleterious effects on plants under extreme conditions, because they can negatively effect on the vital physiological functions and may lead to death, for example, the presence of considerable amounts of toxics in the soil or in the air, deficiency of water, oxygen, or minerals, and extreme conditions for temperature, humidity, oxygen, CO, or light [7].

4. Parasitism and pathogenicity

The term parasitism called on the state in which an organism (parasite) lives on or in another one (host) to obtain its required nutrition. Usually, the parasitism is correlated with pathogenicity, which means the ability of an organism to cause a disease. However, the parasitism in some cases leads to a benefit relationship called symbiosis, in which both plant and organism alternate the benefits, such as bacterial nodules in the roots of legume plants and the mycorrhizal infection of feeder roots of most flowering plants. In the case of parasitism-pathogenicity relationship, the plant is diseased with the appearance of different symptoms such as increased respiration, disintegration or collapse of cells, wilting, abscission, abnormal cell division and enlargement, and degeneration of specific components such as chlorophyll [15]. The most common plant pathogens are fungi, bacteria, mollicutes, parasitic higher plants, parasitic green algae, nematodes, protozoa, viruses, and viroids. These parasites cause serious plant diseases, because they have the ability to penetrate the plant tissues to feed and proliferate in it, and withstand the conditions in which the host lives. These pathogens are also called obligate parasites because they can only live in their living hosts. On the other hand, there are certain pathogens such as most fungi and bacteria can live on either living or dead hosts and on various nutrient media, so they are called nonobligatory parasites. Some nonobligatory parasites can grow saprophytically on dead organic matter, and therefore called semi-biotrophs/facultative saprophytes [16]. There is a type of a life called facultative parasitism, in which an organism grows saprophytically (necrotrophs); however, under certain conditions, they attack living plants and cause a disease; these parasites are called facultative parasites. The type or degree of parasitism does not affect the disease severity. For instance, many diseases caused by weakly parasitic pathogens are much more damaging to a plant than others caused even by obligate parasites. Lysozymes are a main mechanism of most nonobligatory parasites by which they can degrade the plant cell wall and subsequently cause invasion and infection [17].

5. Host range of pathogens

Phytopathogens differ among each other with respect to the plant type, the location of infection, and the age of the organ or tissue (location of infection). The specificity of plant pathogens has various degrees; some pathogens have only one target species of plant, while other pathogens can attack only one genus of plants, and eventually some others have a wide range of hosts, belonging to many families of higher plants. As mentioned above, phytopathogens differ among each other with respect to the location of infection; some of them grow on roots, stems, leaves, fruits or vegetables, and phloem or xylem. Some phytopathogens can only infect the seedlings or the young parts of plants, while the others can only infect the mature tissues [18].

6. Development of disease in plant

The plant disease means the occurrence of physiological disorder(s) due to biotic agents such as microbial infection and/or abiotic agents such as extreme environmental factors. In order for the plant disease to occur, an interaction must happen between two components: the plant and disease cause, which leads to physiological disorders. The disease cause is either biotic agent or abiotic agent as mentioned above. Interestingly, the biotic agents lead to infectious diseases, which develop under suitable environmental conditions. Therefore, the infectious diseases (occurred by pathogens) are not developed under extreme environmental conditions. This means it was impossible to get infectious and noninfectious plant diseases at the same time. The abiotic agents (environmental factors) play an important and vital role in the disease development and severity or disease resistance. This matter depends mainly on different factors: the plant family, the plant age, plant genetic type, pathogen virulence race, pathogen inoculum size, and pathogen dormant state. Therefore, we can imagine the plant disease as a triangle, which is called "disease triangle." The three sides of this triangle are the plant, microorganisms, and the environmental factors. The length of each side is proportional to the sum of the characteristics of the other two sides. For example, if the plant is resistant, the host side and the amount of disease would be small or zero, whereas if the plant is susceptible, the host side would be long and the potential amount of disease could be great [19].

6.1. Disease cycle

The disease cycle is a series of definite events, which lead to the disease development and pathogen propagation. These events include inoculation, prepenetration, penetration, infection, colonization (invasion), and growth and reproduction of the pathogen [20].

6.1.1. Inoculation

Inoculation is the pathogen or any part of the pathogen that contacts with the plant at certain site to initiate the infection process, such as spores, sclerotia, or fragments of mycelium of fungi may be fungal inoculum. In some cases, the inoculum is represented as an intact cell as in bacteria, mollicutes, protozoa, viruses, and viroids. There are two types of inoculum: primary and secondary inoculum, which in turn cause primary and secondary infection. The primary inoculum lives dormant in the winter or summer and causes the original infections in the spring or in the autumn. The secondary inoculum is that produced from primary infections. The primary inoculum is more abundant than secondary inoculum and closer to the crop, and caused more severe diseases and the losses that result. The inoculum has two sources: inside and outside sources. The inside source in which the inoculum is produced on the plant, plant debris, or on the soil, such as fungal and bacterial inocula of perennial plants, is produced on the branches, trunks, or roots of the plants. The outside source of inoculum is in which the inoculum comes into the field with the seed, transplants, tubers, or other propagative organs or it may come from sources outside the field. In some cases, the inoculum is produced on the plant surface as in fungi, bacteria, parasitic higher plants, and nematodes, which either produce their inoculum on the surface of infected plants or their inoculum reaches the plant surface when the infected tissue breaks down. However, the inoculum may be produced within the plant as in viruses, viroids, mollicutes, fastidious bacteria, and protozoa. Interestingly, there is an expression called inoculum landing or inoculum arrival, which means incoming of the inoculum to the host plants passively by wind, water, and insects [21].

6.1.2. Prepenetration

6.1.2.1. Attachment of pathogen to host

Some pathogens directly penetrate the plant tissues by their vectors and then are surrounded by cytoplasm, cell membrane, or cell wall of plant cell, such as mollicutes, fastidious bacteria, protozoa, and most viruses. In other cases, the pathogen firstly makes contact with the external surface of the plant, and then penetration process takes place, such as fungi, bacteria, and parasitic higher plants. The adhesion of the pathogen with plant surface is carried out by mucilaginous substances found on the pathogen surface or at its tip. These substances are composed of mixture of water-insoluble polysaccharides, glycoproteins, lipids, and fibrillary materials, which, when moistened, become sticky and help the pathogen adhere to the plant. In some fungi as powdery mildew, adhesion is carried out by the release of cutinase enzyme from the spore, which makes the plant and spore areas of attachment more hydrophilic and cements the spore to the plant surface [22].

6.1.2.2. Spore germination

Spore germination process initiates by growth stimulation, which takes place with the availability of proper environmental conditions. Once the stimulation has been received, the spore starts to utilize the stored food, such as lipids, polyoles, and carbohydrates to build germ tube as abridge with cell membrane and cell wall of the plant. When appropriate physical and chemical signals, such as surface hardness, hydrophobicity, surface topography, and plant signals, are present, germ tube extension and differentiation take place [23].

6.1.2.3. Appressorium formation and maturation

Appressorium is a specialized cell typical to many fungal plant pathogens that is used to infect the plant host. Once appressoria are formed, they adhere tightly to the leaf surface and then penetrate the plant cell wall via lysozyme secretion [24].

6.1.2.4. Recognition between host and pathogen

When a pathogen comes in contact with a host cell, the plant triggers a signal that either allows or retards the pathogen growth and development of disease. This signal is a biochemical reaction, which acts as a receptor to a pathogen contact. The pathogen propagation depends on the components of the plant cell, such as fatty acids galacturonan, phenolic compounds, strigol, amino acids, and sugars [25].

6.1.2.5. Spores and seed germination

The availability and ability of host infection are increased by vegetative pathogen. The infection by fungal spores or parasitic higher plant seeds is carried out after germination has achieved. Fungal spores' germination is carried out by releasing either a mycelium or a germ tube that grows into the plant cell and cause infection [25].

6.1.2.6. Growth of nematodes

The growth of nematodes starts with hatching of eggs, which essentially requires convenient environmental conditions such as temperature and moisture. After hatching of the eggs, the larvae penetrate the plant cell and grow to form the adults. After maturation, the adults of nematodes closely adhere with plant roots due to some factors, such as carbon dioxide and amino acids [9].

6.1.3. Penetration

Phytopathogens penetrate plant surfaces either through natural openings such as fungi and nematodes or through wounds in cell wall such as bacteria, viruses, viroids, mollicutes, fastidious bacteria, and protozoa. Penetration and infection are not usually correlated together, because some penetrated plants are resistant to phytopathogens [26].

6.1.4. Infection

The intimate contact of phytopathogen with its host is called infection process. The infection process is either successful or unsuccessful depending on the type of host, whether susceptible or resistant, respectively. Successful infection results in the appearance of symptoms, such as discoloration, necrosis, dwarfism, and so on of the host. While unsuccessful (latent) infection does not lead to any observations for the symptoms. As well known, the symptoms start to appear after the incubation period of the pathogen has been finished. The symptoms either are stable or may be changed until death [27].

6.1.5. Invasion

The phytopathogens can invade the plant tissues by producing mycelia which grow between the cuticle and epidermis, such as pathogenic fungi of an apple. Nevertheless, other phytopathogens such as those causing powdery mildews produce mycelia which grow on the plant surface, and then extend to form a structure called haustoria, which in turn extend into the epidermal cells. Therefore, plant pathogenic fungi can invade their host either by intracellular mycelia, which directly grow through the cells, or by intercellular mycelia, which grow between the cells. On the other hand, plant pathogenic bacteria invade the plant tissues via intercellular way, and cause vascular wilts. Although nematodes can invade the plant tissues intercellulary or intracellularly, they usually feed on the epidermal cells by piercing. Other phytopathogens such as viruses, viroids, mollicutes, fastidious bacteria, and protozoa can intracellularly invade the plant tissues [28].

6.1.6. Growth and reproduction of the pathogen

Most phytopathogens especially fungi and parasitic higher plants invade and infect plant tissues through the point of inoculation. Therefore, these pathogens can easily grow and spread within the plant tissues until a certain limit or death occurs. For example, fungi can invade and infect the plant tissue to cause vascular wilts. This invasion is carried out by releasing spores within the vessels [29].

http://dx.doi.org/10.5772/65325

Author details

Waleed M. Abdulkhair[1*] and Mousa A. Alghuthaymi[2]

*Address all correspondence to: waleed_hamada@yahoo.com

1 General Department of Basic Medical Sciences, Microbiology Department, National Organization for Drug Control and Research (NODCAR), Giza, Egypt

2 Biology Department, Science and Humanities College, Shaqra University, Al-Quway'iyah, Saudi Arabia

References

[1] Bruehl GW. Plant pathology, a changing profession in a changing world. Annu Rev Phytopathol. 1991; 29: 1–12.

[2] Cramer HH. Plant protection and crop production. 5th ed. Farbenfabriken Bayer AG; 1967 Leverkusen.

[3] Brown JF, Ogle HJ, Plant pathogens and plant diseases. 3rd ed. Rockvale Publications; 1997 Armidale, Australia.

[4] Daly JM. The role of recognition in plant disease. Annu Rev Phytopathol. 1984; 22: 273–307.

[5] Horsfall JG, Cowling EB. Plant disease. 3rd ed. Academic Press; 1980 New York, NY.

[6] Duncan JM, Torrance L. Techniques for the rapid detection of plant pathogens. 2nd ed. Blackwell; 1992 Oxford.

[7] Keen NT. A century of plant pathology: A retrospective view on understanding host–parasite interactions. Annu Rev Phytopathol. 2000; 38: 31–48.

[8] Singh US, Singh RP, Kohmoto K. Pathogenesis and host specificity in plant diseases: histochemical, biochemical, genetic and molecular bases. 2nd ed. Pergamon/Elsevier; 1995 Tarrytown, NY.

[9] Cobb NA. Contributions to a science of nematology. Pt. 1914; 1: 1–33.

[10] Bridge PD, Couteaudier Y, Clarkson JM. Molecular variability of fungal pathogens. 4th ed. CAB International; 1998 Wallingford, UK.

[11] Kosuge T, Nester EW. Plant-microbe interactions: molecular and genetic perspectives. 3rd ed. Macmillan; 1984 New York, NY.

[12] Goto M. Fundamentals of bacterial plant pathology. 2nd ed. Academic Press; 1993 San Diego.

[13] Bawden FC, Pirie NW, Bernal JD, Fankuchen I. Liquid crystalline substances from virus-infected plants. Nature. 1963; 138: 1051–1052.

[14] Alexopoulos CJ, Mims CW, Blackwell M. Introductory mycology. 4th ed. Wiley; 1996 New York, NY.

[15] Brown W. Studies in the physiology of parasitism. I. The action of Botrytis cinerea. Ann Bot. 1915; 29: 313–348.

[16] Ellingboe AH. Inoculum production and infection by foliage pathogens. Annu Rev Phytopathol. 1968; 6: 317–330.

[17] Dollet M. Plant diseases caused by flagellate protozoa (*Phytomonas*). Annu Rev Phytopathol. 1984; 22: 115–132.

[18] Doi Y, Teranaka M, Yora K, Asuyama H. Mycoplasma or PLT group-like microorganisms found in the phloem elements of plants infected with mulberry dwarf, potato witches broom, aster yellows, or paulownia witches broom. Ann Phytopathol Soc Jpn. 1967; 33: 259–266.

[19] Hancock JG, Huisman OC. Nutrient movement in host pathogen systems. Annu Rev Phytopathol. 1981; 19: 309–331.

[20] Dixon RA, Harrison MJ, Lamb CJ. Early events in the activation of plant defense responses. Annu Rev Phytopathol. 1994; 32: 479–501.

[21] Holmes FO. Local lesions in tobacco mosaic. Bot Gaz. 1929; 87: 39–55.

[22] Romantschuk M. Attachment of plant pathogenic bacteria to plant surfaces. Annu Rev Phytopathol. 1992; 30: 225–243.

[23] Stanley MS, et al. Inhibition of fungal spore adhesion by zosteric acid as the basis for a novel, nontoxic crop protection technology. Phytopathology. 2002; 92: 378–383.

[24] Emmett RW, Parbery DG. Appressoria. Annu Rev Phytopathol. 1975; 13: 147–167.

[25] Hillocks RJ, Waller JM. Soil borne diseases of tropical crops. 5th ed. CABI; 1998 Wallingford, UK.

[26] Nielsen KA, et al. First touch: an immediate response to surface recognition in conidia of *Blumeria graminis*. Physiol Mol Plant Pathol. 2000; 56: 63–70.

[27] Leong SA, Allen C, Triplett EW. Biology of plant-microbe interactions. 1st ed. APS Press; 1984 St. Paul, MN.

[28] Perfect SE, Green JR. Infection structures of biotrophic and hemibiotrophic fungal plant pathogens. Mol Plant Pathol. 2001; 2: 101–108.

[29] Meredith DS. Significance of spore release and dispersal mechanisms in plant disease epidemiology. Annu Rev Phytopathol. 1973; 11: 313–342.

Chapter 5

Making Soil More Accessible to Plants: The Case of Plant Growth Promoting Rhizobacteria

Metin Turan, Nurgül Kıtır, Ülker Alkaya,
Adem Günes, Şefik Tüfenkçi, Ertan Yıldırım and
Emrah Nikerel

Additional information is available at the end of the chapter

http://dx.doi.org/10.5772/64826

Abstract

Plant Growth Promoting Rhizobacteria (PGPR) are beneficial soil bacteria that can live either symbiotically with plants at rhizosphere or as endophytes living on or inside of the host plants. There are two main mechanisms via PGPR contribute to the plant growth. Direct mechanism consists of phytohormone production (i.e. auxins (IAA), cytokinins and gibberellins), biological nitrogen fixation, solubilizing inorganic phosphates, mineralizing organic phosphate and producing organic matter such as amino acids. As indirect mechanisms, PGPR aid plants in combat against the pathogen microorganisms by means of stimulating the disease-resistance mechanism of plants, promote favorable symbiosis, decontaminate the soil of xenobiotics. PGPR can also help plants to cope against abiotic stress by lowering ethylene levels, or against pathogenic microorganism by means of secreting antibacterial/antifungal substances. Exact mechanisms of PGPR characteristics which stimulate the plant growth or product formation are still under investigation, yet in agriculture, PGPR are used as environmental friendly biofertilizers, biocontrol agents or biostimulants. These beneficial bacteria are usually introduced to the plants either in powder or liquid form or the seeds are covered with the inoculants before sowing. Plants are subject to many different environmental elements. Abiotic factors such as drought or water stress have been one of the main plant growth limiting factors. Agricultural PGPR application is an alternative solution against loss due to the environmental stresses, since breeding a plant with stress resistance trait is a very long and tricky process due to the fact that such traits are controlled by multiple genes. PGPR phytohormone and enzyme (i.e. ACC deaminase) production can decrease the stress levels of plants while enhancing the root structures.

Keywords: plant growth–promoting rhizobacteria, enzymes, plant nutrient use efficiency, nitrogen fixation, phosphorus solubilizing, plant stress

1. Introduction

Soil is a composed of minerals, organic matters, water, and microorganisms and it covers the surface of the earth. Soil not only provides an attachment surface for plants but also the necessary materials for their growth. It also acts as host to many types of bacteria. The number of bacterial species living in the soil varies according to the environmental conditions such as temperature of the soil, amount of salt, chemicals, and moisture in the soil, and plants growing nearby in the soil [1]. Bacteria are usually found abundant around the rhizosphere. The term "rhizosphere" was first coined by Lorenz Hiltner in 1904 to define the layer of soil around the plant root that is populated by microorganisms. The relationship between the plant and the soil bacteria can be beneficial, harmful, or neutral according to the environmental conditions surrounding the plant [2]. For example, bacterial species that has the trait to increase phosphate solubility of the plant can only be beneficial when the plant is growing on a phosphate-poor soil. When the phosphate is given to the plant as fertilizer, the bacterium species becomes neutral from the plant point of view.

The human population has been increasing rapidly and the industrialization grows accordingly. This contributes to the fact that not only the current food sources would not be enough even every condition would stay the same but also industrialization has very negative effects on the environment, such as decrease in the available land for agriculture available land, global warming, and air and water pollution. New strategic solutions should be addressed to improve agricultural yields and sustainability so that the food requirements for the human population will be met with the lowest environmental impacts. A likely solution can be the use of "plant growth promoting rhizobacteria (PGPR)", soil bacteria which colonize at the rhizosphere of the plants stimulating the plant growth. The term PGPR was first coined by Kloepper and Schroth in 1978. The type of PGPR is directly related to the products exudate by the plant root such as sugars, organic acids, and proteins. Understanding how the plant chooses which type of PGPR would form the microbial community in the rhizosphere would give us insight when choosing PGPR inoculants for increased plant crop yields and is a major scientific issue [3].

2. Mechanisms of PGPR

There are numerous types of bacteria which have been observed to possess at least one PGPR trait such as *Pseudomonas, Azospirillum, Azotobacter, Klebsiella, Enterobacter, Alcaligenes, Arthrobacter, Burkholderia,* and *Bacillus.* These are commercial inoculant species used as biofertilizers which enhance the crop yield, bioprotectors which defend the plant against pathogens and biostimulators which produce phytohormones beneficial to the plant [4]. The

PGPRs are offering a cheaper and more environmental friendly option compared to using chemical pesticides, herbicides, and fertilizers [5]. They affect the plants in couple of different mechanism, yet none of them is well understood to this day.

2.1. Direct mechanisms

2.1.1. Plant growth substances

Phytohormones, another name for plant growth substances, are plant hormones or messengers that influence the plant's response to its environment. These organic compounds are produced in one part of plant in a very low amount and carried into the other locations of the plant [4]. The physical responses gained by these hormones are ripening or growth of roots and leaves.

There are five main types of phytohormones: auxins, gibberellins, ethylene, cytokinins, and abscisic acid. PGPR usually produces cytokinins, gibberellins, and IAA as phytohormones.

Cytokinins are compounds whose structure is similar to adenine. As the name suggests this hormone induces cytokinesis (cell division) in plants thus involves in growth, root initiation, increase in root surface area [6, 7]. This hormone can be synthesized by plant, some PGPRs, and yeast strains [8]. Some phytopathogens are also reported to synthesize cytokinins, but the amount of the produced hormone regulates whether it promotes or induces plant growth. Various bacterial strains of *Azotobacter* spp., *Rhizobium* spp., *Pantoea agglomerans*, *Rhodospirillum rubrum*, *Pseudomonas fluorescens*, *Bacillus subtilis*, and *Paenibacillus polymyxa* are recorded to produce cytokinins [9].

Gibberellin, another phytohormone is also synthesized by some cytokinin-producing PGPR. Gibberellin has a role in flowering, germination, dormancy, sex expression, and plant growth. The gibberellin and cytokinin mechanisms for bacterial production and regulations are now fully understood. Thus, the known effects of these hormones come from the plant physiological knowledge [10].

IAA (indole-3-acetic acid) is the most significant and most studied auxin produced by plants [9] and PGPR which has role in cellular responses such as cell division, organogenesis, gene expression, pigment formation, seed germination, root development, photosynthesis, and tropic responses (such as to gravity and light) [9, 11]. IAA also has role in stress resistance of plants [12]. Like cytokinin, the amount of IAA can be both inhibitory and stimulatory. The amount of IAA that required for the plant growth promotion is influenced by the plant species and the bacterial species [4]. Since IAA is responsible for root formation and lengthening, one of the effects of IAA on plants is increasing the amount of nutrients by the root and the amount of exudation from the root [13]. The increase in exudations promotes the increase in biomass of PGPR and the nodule formation in the rhizosphere [14].

Phytohormone ethylene is responsible for ripening of fruit, promoting root growth, activation other phytohormones, inhibiting formation of *Rhizobia* spp. nodule formations. It is also synthesized when the plant is faced with biotic and abiotic stresses [15].

2.1.2. Biological nitrogen fixation

Nitrogen is an essential element for life as it is present in the structures of important biochemicals such as proteins and nucleotides. Although the air is rich with N_2 (g), plants, and many other complex organisms cannot use nitrogen in this form. Biological nitrogen fixation (BNF) process by N-fixing bacteria produces the ammonia which can be used by plants as a nitrogen source. Plants biomass and product yields are limited by the amount of nitrogen available, thus applications of N-containing fertilizers is heavily used in agriculture. The downside of using chemical fertilizers is they are expensive and have negative impact on environment. Using PGPR and providing needed nitrogen by the BNF can be an alternative way to increase agricultural yield [16, 17].

Biological nitrogen fixation fortunately not limited to the PGPR that forms symbiotic nodules with legumes, but there are nonsymbiotic free living nitrogen fixing bacteria as well. *Azospirillum, Azoarcus, Azotobacter, Bacillus polymyxa, Burkholderia, Gluconoacetobacter,* or *Herbaspirillum* are such bacterial species reported to have PGPR properties [18].

2.1.3. Phosphate solubilizing bacteria (PSB)

Nitrogen is not the only crucial element for life which can limit the plant growth. For example, phosphorus is also essential for the plants. Soil holds large amounts of phosphate, yet it is found in insoluble form. Some PGPR are reported to solubilize the phosphate in the soil through acidification, chelation, or enzymatically [19] *Gluconacetobacter diazotrophicus* is a PGPR native to sugarcane that has the property to solubilize phosphate via acidification [20].

2.2. Indirect mechanisms

2.2.1. Biocontrol via antibiotics and lytic enzymes

It has been known that, microorganisms compete against each other for nutrients, colonization sites in their natural environments. Many PGPR species evolved mechanism to reduce competition such as releasing of antibiotics, lytic enzymes, or weak organic acids to the environment. This characteristic of PGPR makes it a valuable tool against plant pathogens [18]. Yet, increase in the usage of antibiotic producing bacteria might result in development of resistant pathogens.

The enzymes that PGPR secrete to eliminate pathogens such as *Botrytis cinerea, Sclerotium rolfsii, Fusarium oxysporum, Phytophthora* spp., *Rhizoctonia solani,* and *Pythium ultimum* are chitinases, cellulates, proteases, and lipases which can destroy the cell walls of the pathogens [9].

2.2.2. Induced systemic resistance (ISR)

Induced systemic and systemic acquired resistances are response mechanisms that plants evolved against pathogens. Unlike systemic acquired resistance (SAR) which is triggered by infection by a pathogen, in ISR, the trigger is a PGPR which will make the plant resistance to phytopathogens. ISR starts at the root and spreads to the shoots [21]. This phenomenon was first observed in 1991 by van Peer et al. They infected *Arabidopsis thaliana* plant root with nonpathogenic *Pseudomonas* spp. and found out the rest of the plant also gained resistance to pathogenic bacteria. Since this discovery, ISR has been studied in many plants such as bean, tobacco, and tomato [18]. Plants with ISR response react to pathogenic bacteria faster and stronger. It should be also noted that ISR response is not pathogen-specific and can be used to stimulate plant immune response against more than one pathogen species [9].

2.2.3. Siderophore production

Iron is another essential nutrient for plants. In aerobic conditions, iron is found as Fe^{3+} form which is not soluble for microorganisms and plants. Some microorganisms produce and secrete low mass iron chelators. These chelators are called siderophores and have high affinity for iron. These operate as solubilizing agents for Fe^{3+} in limiting conditions. Fe^{3+} becomes Fe^{2+} for while entering the cell membrane and then unbind from the siderophores inside the cell [22].

Siderophore production is also observed to be a biocontrol mechanism, since with this process, PGPR derives other microorganisms from iron. PGPR also reported to use siderophores to obtain other heavy metals (such as arsenic) from the soil and prevents the heavy metal toxicity in plants [23]. This characterization can be used for bioremediation of the heavy metal toxic soil as well.

2.2.4. Regulation of stress conditions

Ethylene is a phytohormone which is also secreted as response to biotic and abiotic stresses such from salt, drought, or pathogenic bacteria. Although promoting growth and ripening of fruits, in high amounts ethylene have harmful effects on the plant. Many PGPRs synthesis an enzyme called ACC deaminase, which destroys the precursor of ethylene called 1-aminocyclopropane-1-carboxylate (ACC), thus decreasing the ethylene levels and relieving the stress of the plant [24].

Some PGPRs which do not have the ability to produce ACC deaminase, can also promote the growth of plants via secretin of IAA even though other inhibitory factors are found in the environment [9].

3. Examples of PGPR

3.1. Symbiotic PGPR: rhizobacteria

Rhizobacteria are soil bacteria which colonize at the root of legumes forming nodules. They fix the atmospheric nitrogen for the plant benefit in exchange for carbon source. Rhizobacteria are the most known PGPRs. Inoculation with rhizobacteria provides biomass increase in legumes [4, 9].

Rhizobacteria are host-specific bacteria, meaning that they will not form rhizosphere nodules with any type of plants. The most common rhizobacteria are Rhizobium and Bradyrhizobium. They are both Gram-negative, rod-shaped (bacilli) bacteria. Rhizobium forms symbiotic nodule with vetches, peas, lentil, clovers, and beans [4].

3.2. Nonsymbiotic PGPR

Fortunately, nitrogen fixation is not limited to Rhizobacteria. There are many free-living species which can also perform biological nitrogen fixation.

Some important nonsymbiotic nitrogen-fixing bacteria include *Azoarcus* sp., *Gluconacetobacter diazotrophicus, Herbaspirillium* sp., *Azotobacter* sp., *Achromobacter, Acetobacter, Alcaligenes, Arthrobacter, Azospirillum, Azomonas, Bacillus, Beijerinckia, Clostridium, Corynebacterium, Derxia, Enterobacter, Klebsiella, Pseudomonas, Rhodospirillum, RhodoPseudomonas,* and *Xanthobacter* [25].

Applications of *Azotobacter* and *Azospirillum* species are reported to increase yield of grass type of crops. Although Azospirillum has been isolated from cereal initially, it has been used to inoculated noncereal crops more frequently. It is stated that Azospirillum bacteria is not a host-specific species but a general root colonizer [4].

The family *Acetobacteriaceae* includes genera, *Acetobacter, Gluconobacter, Gluconoacetobacter,* and *Acidomonas. Gluconacetobacter diazotrophicus* is an acid tolerant, Gram-negative and obligate aerobe bacteria. The bacteria cells can grow on high sucrose content and low acidity. The optimum sugar concentration is 10% and pH is 5.5 for growth although its recorded that they can live at pH 3. It is an endophyte; located at the internal tissues of its host [16].

4. Conclusion

Plant growth promoting rhizobacteria are being intensify researched to increase crop yields, to protect the plants and stimulate the plant growth via phyhormone production. Even though the mechanisms behind PGPR characteristics are not fully discovered, there are many commercialized PGPR strains which are used as agricultural inoculants. These strains are *Agrobacterium radiobacter, Azospirillum brasilense, Azospirillum lipoferum, Azotobacter chroococcum, Bacillus licheniformis, Bacillus megaterium, Bacillus mucilaginous, Bacillus pumilus, Bacillus* spp., *Bacillus subtilis, Bacillus subtilis* var. *amyloliquefaciens, Burkholderia cepacia, Paenobacillus macerans, Pantoea agglomerans, Pseudomonas aureofaciens, Pseudomonas chlororaphis, Pseudomonas solanacearum, Pseudomonas* spp., *Pseudomonas syringae, Serratiaentomophilia, Streptomyces griseoviridis, Streptomyces* spp., *Streptomyces lydicus,* and various *Rhizobia* spp.

The inoculation of agricultural plants with PGPR still makes the minor fraction of crop enhancement methods. To increase the application of PGPRs, the mechanisms that are unknown should be studied, the differences and advantages of using nonsymbiotic PGPR over rhizobacteria species should be determined. The production and storage of the PGPR inoculants should be addressed.

Due to the climate and soil composition, there are inconsistencies between the greenhouse trials' results and the field trials should be minimized. A computational approach can be used to find the interaction of plant and PGPR and simulate physiological responses under certain environmental conditions.

Author details

Metin Turan[1*], Nurgül Kıtır[1], Ülker Alkaya[1], Adem Günes[2], Şefik Tüfenkçi[3], Ertan Yıldırım[4] and Emrah Nikerel[1]

*Address all correspondence to: m_turan25@hotmail.com

1 Yeditepe University, Engineering Faculty, Genetics and Bioengineering Department, Istanbul, Turkey

2 Agricultural Faculty, Soil and Plant Nutrition Science, Erciyes University, Kayseri, Turkey

3 Yüzüncü Yıl University, Soil and Plant Nutrition Science, Van, Turkey

4 Ataturk University, Agricultural Faculty, Horticultural Department, Erzurum, Turkey

References

[1] Glick BR, Patten CL, Holguin G, Penrose DM. Biochemical and Genetic Mechanisms Used by Plant Growth Promoting Bacteria. London, UK: Imperial College Press; 1999.

[2] Lynch JM, editor. The Rhizosphere.Chichester, UK: Wiley-Interscience; 1990.

[3] Drogue B, Doré H, Borland S, Wisniewski-Dyé F, Prigent-Combaret C. Which specificity in cooperation between phytostimulating rhizobacteria and plants? Res Microbiol. 2012; 163: 500–510.

[4] Saharan BS. Plant growth promoting rhizobacteria: a critical review. Life Sci Med Res. 2012.

[5] Stefan M, Mihasan M, Dunca S. Plant growth promoting Rhizobacteria can inhibit the in vitro germination of *Glycine max* L. seeds. Scientific Ann. Univ. "Alexandru Ioan Cuza" Iasi. Sect Genet Mol Biol, T. IX, 2008; 3: 105–110.

[6] Salamone IEG, Hynes RK, Nelson LM. Role of cytokinins in plant growth promotion by rhizosphere bacteria. In: Siddiqui ZA, editor. PGPR: Biocontrol and Biofertilization. The Netherlands: Springer; 2005. vol. 173. p. 95.

[7] Werner T, Motyka V, Laucou V, Smets R, Van Onckelen H, Schmülling T. Cytokinin-deficient transgenic Arabidopsis plants show multiple developmental alteration

syndicating opposite functions of cytokinins in the regulation of shoot and root meristem activity. Plant Cell. 2003; 15: 2532–2550.

[8] Sakakibara H. Cytokinins: activity, biosynthesis, and translocation. Annu Rev Plant Biol. 2006; 57: 431–449.

[9] Glick BR. Plant growth-promoting bacteria: mechanisms and applications. Scientifica. 2012.

[10] Vessey JK. Plant growth promoting rhizobacteria as biofertilizers. Plant Soil. 2003; 255: 571–586.

[11] Ashrafuzzaman M, Hossen FA, Ismail MR, Hoque MA, Islam MZ, Shahidullah SM, Meon S. Efficiency of plant growth promoting Rhizobacteria (PGPR) for the enhancement of rice growth. Afr J Biotechnol. 2009; 8: 1247–1252.

[12] Spaepen S, Vanderleyden J. Auxin and plant-microbe interactions. Cold Spring Harbor Perspect Biol. 2011; 3: 4.

[13] Badenoch-Jones J, Summons RE, Rolfe BG, Letham DS. Phytohormones, Rhizobium mutants, and nodulation in legumes. IV. Auxin metabolites in pea root nodules. J Plant Growth Regul. 1984; 3: 23–39.

[14] Mathesius U, Schlaman HRM, Spaink HP. Sautter C. Rolfe B, Djordjevic MA. Auxin transport inhibition precedes root nodule formation in white clover roots and is regulated by thavonoids and derivatives of chitin oligosaccharides. Plant J. 1998; 14: 23–34.

[15] Abeles FB, Morgan PW, Saltveit Jr. ME. Ethylenein Plant Biology. 2nd ed. New York, NY, USA: Academic Press; 1992.

[16] Muthukumarasamy R, Revathi G, Seshadri S, Lakshminarasimhan C. Gluconacetobacter diazotrophicus (syn. *Acetobacter diazotrophicus*), a promising diazotrophic endophyte in tropics. Curr Sci. 2002; 83: 137–145.

[17] Eskin N, Vessey K, Tian L. Research progress and perspectives of nitrogen fixing bacterium, *Gluconacetobacter diazotrophicus,* in monocot plants. Int J Agron. 2014; 1–13. http://doi.org/10.1155/2014/208383.

[18] Pérez-Montaño F, Alías-Villegas C, Bellogín RA, Del Cerro P, Espuny MR, Jiménez-Guerrero I, Cubo T. Plant growth promotion in cereal and leguminous agricultural important plants: from microorganism capacities to crop production. Microbiol Res. 2014; 169: 325–336.

[19] Hameeda B, Harini G, Rupela OP, Wani SP, Reddy G. Growth promotion of maize byphosphate solubilizing bacteria isolated from compost and microfauna. Microbiol Res. 2008; 163: 234–242.

[20] Bertalan M, Albano R, de Pádua V, Rouws L, Rojas C, Hemerly A, Ferreira PC. Complete genome sequence of the sugarcane nitrogen-fixing endophyte *Gluconacetobacter*

diazotrophicus Pal5. BMC Genomics. 2009; 10: 450. http://doi.org/10.1186/1471-2164-10-450.

[21] Ramos SB, Barriuso MJ, Pereyra IMT, Domenech J, Gutiérrez MFJ. Systemic disease protection elicited by plant growth promoting rhizobacteria strains: relationship between metabolic responses, systemic disease protection, and biotic elicitors. Phytopathology. 2008; 98: 451–457.

[22] Boukhalfa H, Crumbliss AL. Chemical aspects of siderophore mediated iron transport. Biometals. 2002; 15: 325–391.

[23] Wang J, Zhao FJ, Meharg AA, Raab A, Feldmann J, McGrath SP. Mechanisms of arsenic hyperaccumulation in *Pteris vittata*. Uptake kinetics, interactions with phosphate, and arsenic speciation. Plant Physiol. 2002; 130: 1552–1561.

[24] Glick BR, Cheng Z, Czarny J, Duan J. Promotion of plant growth by ACC deaminase-producing soil bacteria. Eur J Plant Pathol. 2007; 119: 329–339.

[25] Saxena AK, Tilak KVBR. Free-living nitrogen fixers: Its role in crop production. In: Verma AK, editor. Microbes for Health, Wealth and Sustainable Environment. New Delhi: Malhotra Publ. Co.; 1998. p. 25–64.

Section 2

Physiology

Chapter 6

The Unsuspected Intrinsic Property of Melanin to Transform Light into Chemical Energy and the Seed Growth

Arturo Solís Herrera

Additional information is available at the end of the chapter

http://dx.doi.org/10.5772/64542

Abstract

Seeds are very important part of the world's diet, contributing to half of the global per capita energy intake. Thereby, their study has a substantive relevance, reflected by numerous yearly publications. However, mysteries remain about the main molecular mechanisms involved in germination and dormancy. Seed is a completely independent living thing, and in suitable conditions, hatches and generates a new adult plant completely, identical to which they gave rise. And to do so requires only light and water in certain proportions. Theoretically, the seed has reserves of nutrients that allow it to grow, until their so-called autotrophic features allow them to establish itself as a self-sufficient organism. So far, the above cannot be explained adequately, we only have abundant theories that come and go. However, our finding of the intrinsic property of melanin is that it transforms the visible and invisible light to chemical energy through the water molecule dissociation and marks a before and an after process in the study of the germination of the seeds. Nutrients that can be found in a seed not only provide energy but also elements to be biomass, that is, mainly carbon chains of different lengths and combinations, which eventually constitute the backbone of more than 95% of biomolecules. The chemical energy that the seed requires to carry out the highly complex chemical reactions necessary for hatching is taken from water, dissociating it through melanin.

Keywords: melanin, chlorophyll, water dissociation, dormancy, germination, energy

1. Introduction

Life in plant is cyclical, like in human beings. The seed in plants can be considered the starting point; thereby, seed biology is one of the most extensively researched areas in plants physiology. However, two fundamental questions remain: how does the embryo emerge and how is embryo emergence blocked?

It seems not be advantageous for a seed to germinate freely. Thereby, seed dormancy is a mechanism that optimizes the germination. However, little progress has been made toward the understanding of dormancy, besides we do not know the defining events in germination.

The threshold stimulus that initiates a common signal transduction cascade that coordinates diverse cellular responses is unknown and varies widely among individual seeds. Characteristically, germination commence with the uptake of water by the quiescent dry seed. The metabolic activities of the imbibed dormant seeds are only subtle different from those of nondormant seed. The main difference is that the radicle (the embryonic axis) fails to elongate.

Dormancy phenomenon has no clear definition, and it differs among species. When the embryo is constrained by surrounding structures, the phenomenon is known as coat-enhanced dormancy. Interestingly, embryos isolated from these seeds are not dormant. So far, dormancy is a poorly understood phenomenon.

The release of dormancy and completion of germination occur within a relatively few cells associated with the embryonic root axis. However, even the presence of apparently nonresponding cells in the axis and other seed parts helps in a successful germination.

The release of dormancy can be triggered by a variety of known and unknown environmental and chemical stimuli, but the presence of water is unavoidable. Initially, the uptake of water by a mature dry seed is rapid, followed by a plateau phase. A further increase in water uptake occurs only after germination is completed.

The influx of water shows peculiarities as the water inside the cell is pure; thereafter, the dormant cell membranes have an immediate and rapid leakage of solutes and low molecular weight metabolites into the surrounding imbibition solution; however, the processes initiated by the entrance of surprisingly clean water have a certain order, and they are not at random at all and have a definite sequence.

The membranes return to their more stable configuration after a short time of rehydration, and solute leakage is curtailed. Therefore, it is not by chance that it occurs, instead, a highly ordered sequence of events happens; for instance, the amount of *N*-acetyl phosphatidylethanolamine, a phospholipid compound with membrane-stabilizing properties, increases as does that of the corresponding synthase.

Upon imbibition by this pure water, the quiescent dry seed resumes metabolic activity. It is hard to explain how the sole reintroduction of clean water during imbibition is sufficient for metabolic activities to resume, and in matter of hours, a full metabolic status is achieved.

The word "metabolism", which means continuous change, is the result of different processes that occur in a very coordinated manner, both in space as well as in time and place. In no way are the results by chance. And to this we add the energy required so that metabolism can occur, then we try to find a source of energy which is capable not only of producing changes, but also in a proper, orderly, and complex fashion. We can get the answer in the melanin.

http://dx.doi.org/10.5772/64542

2. Melanin, the great transducer

The so-called germination comprises a chain of concatenated events, and by no means they are isolated and all are connected among themselves in a way that we do not understand. But the unsuspected intrinsic capacity of melanin becomes the visible and invisible light energy chemical through dissociation and subsequent reformation of the water molecule. Discovered by our team in 2002, it comes to fill a very important gap in the knowledge of seed biology.

So far, melanin has been considered as the perfect protection against UV-induced photodamage [1]. It is an absorbent filter that reduces the penetration of UV through the epidermis. Other significant properties of melanin already described in the literature are its functions as a free radical scavenger and superoxide dismutase that reduces reactive oxygen species (ROS) [2].

Melanin, under certain circumstances, can also have toxic properties [3], at least theoretically. However, more than 120 genes have been shown to regulate pigmentation in mammals. On the other hand, melanin's basic structure is poorly defined. Melanins are electron acceptors and charge exchange is considered a major binding force in many reactions.

Much has been written about complex properties of melanin, but the mechanisms of action were more theoretical than real, since some properties of melanin seem to oppose each other. But the hitherto unknown property of melanin transforming light into chemical energy breaks the paradigm in which it corresponds to their biological function.

Energy production is the new feature that we now know in melanin, and it is the same everywhere, for example, on the outer layers of the seeds (episperma) and fruits (**Figures 1–9**).

Figure 1. Melanin is present in all the seeds. The difference in hue depends on the concentration of the molecule, granule size, orientation in the molecule, and type of structures that surround it. Photograph shows the high concentration on mature fruit of avocado (American Persea Mill) peel, even it seems that higher concentration of melanin results in greater fruit and seed size, as it is conceivable that the higher amount of melanin means greater availability of chemical energy. On the right side of the picture, we observe tamarind seed (*Tamarindus indica*) whose brown color is also due to the melanin.

Figure 2. Melanin must be present in all cells. Because of it, cells are capable of producing their own chemical energy through dissociation of water molecule, which appears as a universal mechanism. Melanin is concentrated in the areas of increased demand for power, for example, around the nuclei of the cells, and in seeds, in areas of high metabolic demand (episperm). Typically, melanin tends to be close to light energy sources (i.e., sunshine), and on the other hand, the concentration of melanin also functions as a regulator of the amount of light that should reach the inside of the body; in this case, mature avocado fruit and tamarind seeds.

Figure 3. The superficial part of the outer portion of the mature fruit of the avocado is very dark, because of its high concentration of melanin, which indicates that, on the one hand, it transforms significantly visible and invisible light into chemical energy, and on the other hand, regulates the amount of light that must pass through and reach the inner parts of the fruit. Outer covering of the seed of the avocado (episperma) presents a very similar color, and thereby, similar melanin content to the tamarind seeds.

Figure 4. Outer covering of the seed of tamarind (left) and avocado (right)—episperma—present very similar colorations. The melanin present in these structures transforms the light into chemical energy, and also regulates the amount of light that must pass through and reach the deep parts, which also require light, either visible or invisible.

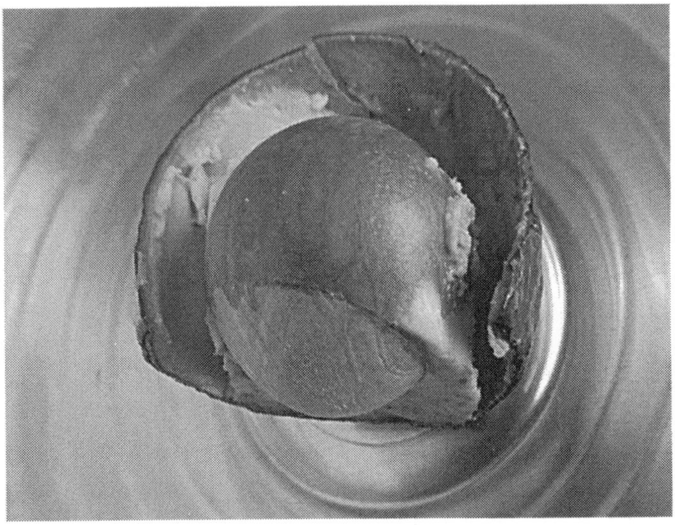

Figure 5. The outer coat of the avocado has melanin of high molecular weight (dark melanin or eumelanin), and the outer coat of the seed (episperma) has melanin of low molecular weight (pheomelanin). In both cases, the function of melanin is the same, chemical energy production.

Figure 6. Eumelanin in the outer casing of the avocado, in addition to producing energy, functions as a regulator of the amount of visible and invisible light that penetrates the inside of the fruit, because it must be within a certain range; on the other hand, the pheomelanin of the seed allows a greater light to pass so that is still maintained within the optimal range.

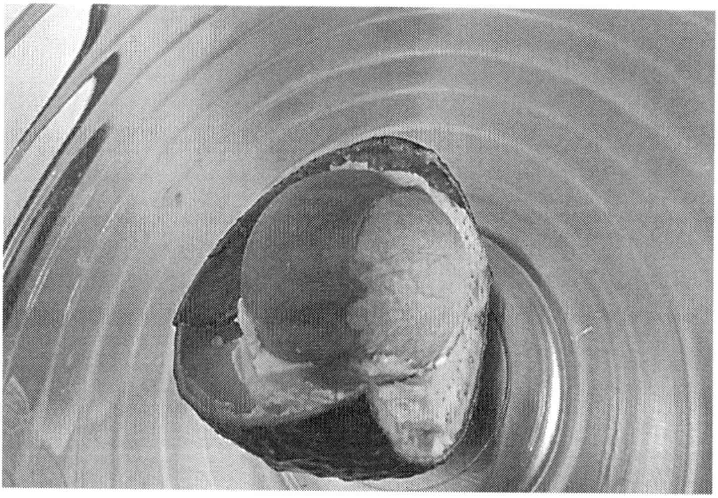

Figure 7. Outer dark color of various brown tones of episperma indicates the presence of melanin. Pulposus green content allows us to infer the presence of heme groups, which are also capable of dissociating water molecule, but unlike the melanin, make it irreversibly.

Figure 8. The grooves that are seen on the surface of the seed, covered and merged in brown tissues (melanin), transport liquids whose nature is not well understood, chemical processes occur inside them even, as in the SAP from the trunk of the tree. But two of the mysteries seem to be solved with melanin, the energy needed to move the fluid and the chemical energy that drives very tidy reactions happening inside these channels, because in both cases, the chemical energy required, no doubt comes from the melanin.

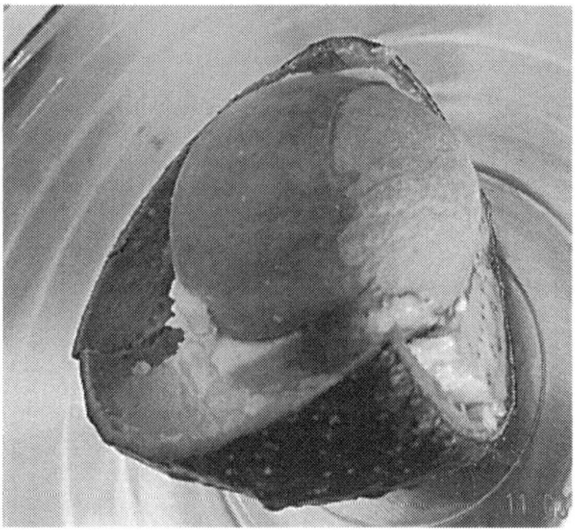

Figure 9. Detail of the groves on the surface of the avocado seed—episperma.

Melanin is able to dissociate the water molecule and also reshapes it. That is, liquid water transforms it into its gaseous components, hydrogen and oxygen, releasing both in molecular form. In this phase, the energy that is released is transported by hydrogen, which is an observed fact in the entire universe. Hydrogen is the main carrier of energy, whereas oxygen is toxic at any concentration.

The belief that our body combines glucose with oxygen to get energy no longer holds since the discovery of diabetes. On the other hand, our body has handled the glucose since the beginning of time, but when the conditions are true, diabetes occurs. When chemical energy levels are not adequate, the body cannot do what it does, which it has been doing since millions of years and millions of times.

In the seed, what happens is that the melanin that is mainly in the outermost layers begins to absorb water and begins to dissociate and re-form, and as a result water that forms inside the seed is pure water because it comes from the recombination of the gases.

But not just the presence of water, but also the levels of hydrogen and oxygen inside the seed begin to rise, and the energy that carries the hydrogen starts to promote each and every chemical reactions that take part in the germination process.

The energy that comes from the melanin is surprisingly consistent and accurate, and hence their effects on seed are also consistent and accurate.

The reaction occurring inside the melanin can be written as follows:

$$2H_2O \leftrightarrow 2H_2 + O_2 + 4e^- \quad (1)$$

For every two molecules of water that is reformed, four high-energy electrons are generated [4].

3. Seed germination, resolved

Therefore, the long-lasting mystery of the germination of the seed seems to be resolved, and they are as follows: when the amount of water and light is adequate, the chemical energy that emanates from the melanin will be enough to promote each and every one of the chemical reactions that occur inside the seed, and this will help in the germination process. But when the amount of water and light is not adequate, the energy that emanates from the melanin will not be enough, and as a result it will only maintain the shape of the seed, but cannot hatch until the levels of chemical energy to the interior of the cell are within a range that can be considered optimal.

4. Is melanin a natural fertilizer?

It is surprising to find nature's insistence to place melanin practically in all the seeds, at different tissues, and different concentrations (**Figures 9–13**). The reason for this is chemical energy production.

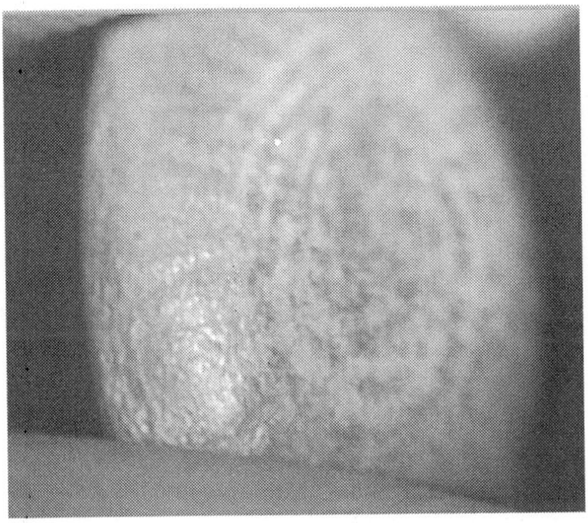

Figure 10. Tamarind seed, 16×. Brown color is given by the melanin.

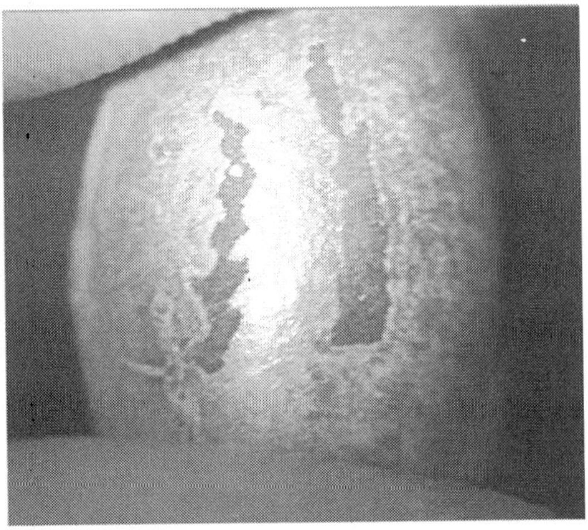

Figure 11. Tamarind seed, 16×. When the external coat is slightly removed a darkest coat of melanin is perceived.

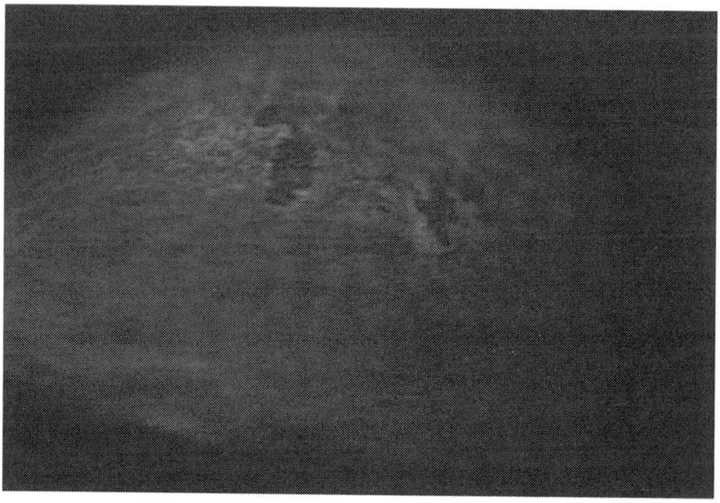

Figure 12. Tamarind seed. With dim light, the melanin is clearly observed and its color shows a resemblance with the eye iris color.

Figure 13. With brightest light, the color of the tamarind seed registered by the camera is almost yellow.

In order to perform its function to generate chemical energy, melanin absorbs visible and invisible light, which is the darkest substance known; and it dissipates the energy by dissociating the water molecule [5], which is reformed almost immediately, starting a very consistent and reliable cycle, while both light and water are still available. The energy that is released due to water molecule breakdown is transported by the diatomic hydrogen.

If we could see how hydrogen bubbles emerge from the melanin, one could observe symmetrically formed and drawn hydrogen bubbles in all the directions (**Figure 14**).

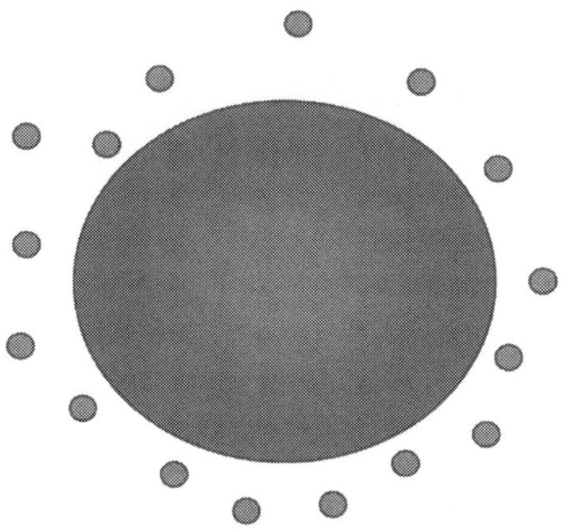

Figure 14. Melanin in dark, and molecular hydrogen in blue, drawing as small bubbles.

Something similar happens in seeds too. However, as melanin is arranged in layers, that is, in episperma, the bubbles displacement occurs outside and inside of the seed (**Figure 15**).

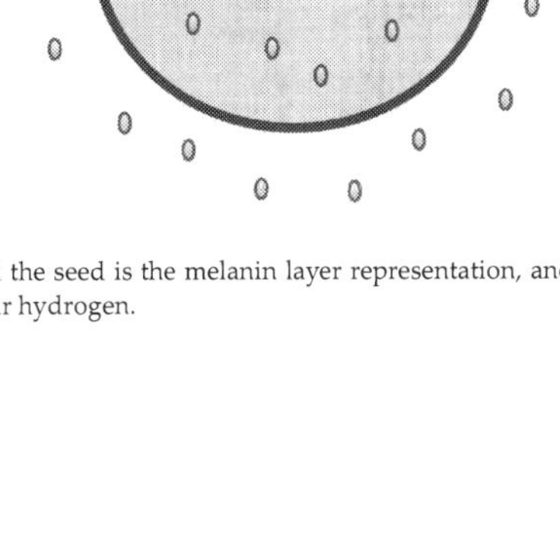

Figure 15. The dark circle around the seed is the melanin layer representation, and the bubbles of blue pale color are the representation of the molecular hydrogen.

The hydrogen that penetrates the inside of the seed has the important function of activating the complex biochemical machinery of germination; driving each of the chemical reactions involved in precise, consistent, and relentless form.

However, the hydrogen bubbles that go into surroundings of the seeds carry out an important function, that is, promote changes in the immediate environment of the seed, which in turn promotes germination, that is, the development of other forms of life, for instance, fungi (**Figures 16–19**).

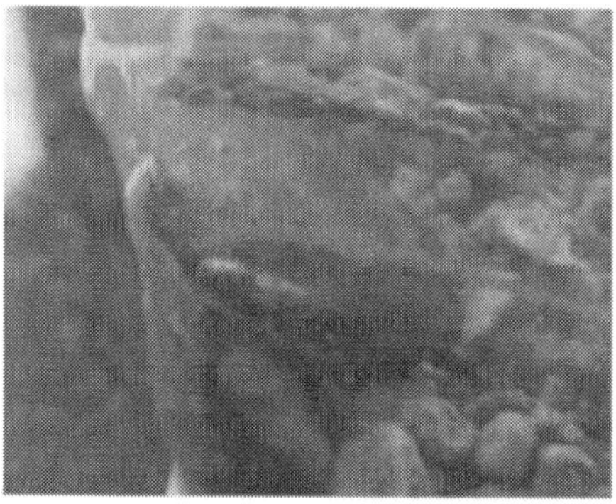

Figure 16. The seed of the tamarind, when not in adequate conditions of water and light, remained almost unchanged, retaining a surface similar to the stored seeds.

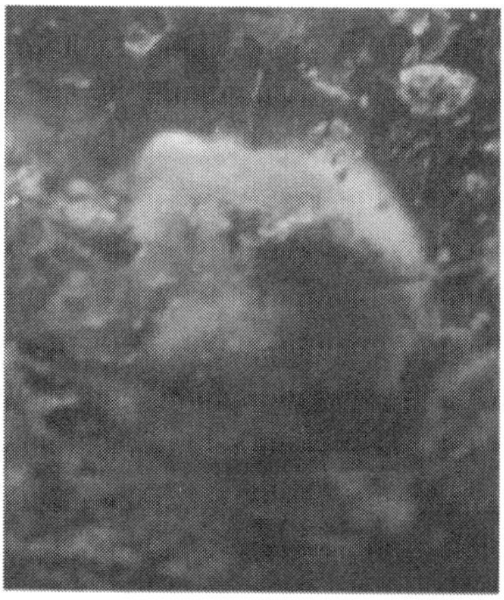

Figure 17. However, when the tamarind seed is in right conditions, by principle of accounts, straightening (polarity) immediately increases its volume, and begins to cover a milkweed (fungi), which also form a part of the cycle (Rhizome).

http://dx.doi.org/10.5772/64542

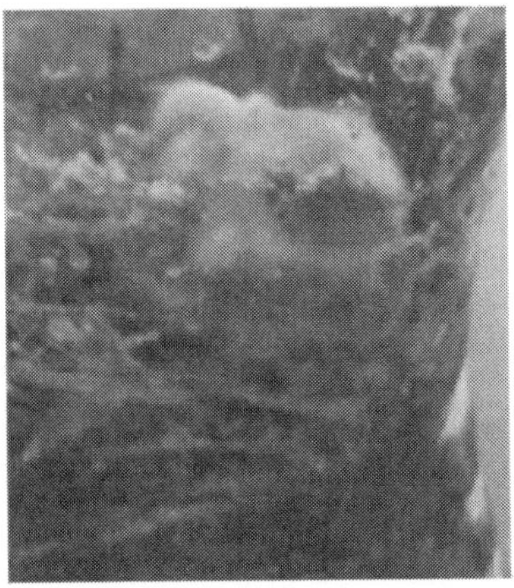

Figure 18. In this photograph, the elongated epicotyl is more visible. Tamarind has hypogeal germination.

Figure 19. In this other specimen, the elongated epicotyl forming the hook is clearly visible.

According to Darwin, although there are more than one million species, there is just one life. For us it means that life originates from the dissociation of the water molecule. Molecular hydrogen being the important part as it is the carrier of energy for excellence in nature, not only on Earth but throughout the universe, and therefore the seeds are no different.

The formation of symmetrical hydrogen bubbles in all the directions by melanin (**Figure 14**) produces a high energy area in its immediate surroundings, so it activates not only the mysterious metabolic processes involved in germination, but produces a beneficial effect on microorganisms that live in the soil, which also play a substantive role in the cycle.

In addition to being the carrier of energy most used in nature, molecular hydrogen is the best antioxidant that is known, because it is able to reduce oxygen into water.

The growth of fungi on the surface of the seed (**Figure 17**) is a clear example of the effect it has on life (in all its forms). The molecular hydrogen and its precious cargo of energy that was injected at the time of the dissociation of the water molecule is very useful, as the seed uses chemical energy in many ways. And it is not any kind of energy, because the melanin generates it in a very precise, consistent, and relentless form, and it occurs both day and night. And this is also reflected in the metabolic processes of the plant, which also are precise, consistent, and incessant, and they occur both day and night.

It was simply the chemical energy with which the seed was created over billions of years of evolution. If we compare the chemical energy that emanates from the melanin with another type of energy, for example, heat, which is a very messy form of energy, germination does not occur.

Even if we compare the energy that comes from the melanin with the energy of the Sun, germination cannot take place from the energy of the Sun. Germination needs both visible and invisible light, which is "conditioned" to transform by a transductor. And in the case of life, melanin seems to be the one which carries out such adaptation of solar energy. As when light energy is too strong, melanin decreases it, and when it is scarce, melanin elevates it. Seems that melanin thrives on pressure.

Melanin is therefore a natural fertilizer, which not only fertilizes the seed itself, but also the nearby environment that surrounds it, optimizing the possibilities of germination.

5. Conclusion

Any chemical process that is subject to a consistent and accurate energy will result in consistent and accurate products.

Life is a continuous process and atoms which are influenced by the melanin begins to behave in a way *sui generis*, and the molecules that are formed and continue to receive chemical energy that emanates from melanin also behave in a very different way than as they would under other conditions, for example, with another type of energy—heat.

And the same might be said for each and every one of the molecules that are continuously formed and grow, becoming more and more complex. And surprisingly, the fundamental chemical energy remains the same, as this is not created or destroyed. But the complexity of the compounds that are formed under the influence of melanin appears to be exponential and at the same time surprisingly exact, because it is repeated over and over again and over millions of years.

Seeds, considered for some authors as mysterious genetic capsules [6], contain many secrets that are yet to be revealed. Mechanisms by which seeds undergo extreme desiccation without losing viability remain unclear. Further, the network of genes conferring on seeds the ability to remain dormant has yet to be fully elucidated.

As the global population exceeds more than 7 billion, and seeds providing more than 70% of the world's caloric intake, the efficient production of seeds is becoming ever more important. Thus, the determination of the genetic and biochemistry of seeds, with the main aim of enhancing yield and nutritive values of seeds, are essential steps. Recently, the utilization of seeds as a source of biofuels to replace fossil fuels in developed countries significantly worsens the problem of adequate food production.

While almost every year new genomic sequences from plants are published, the impact on crops has not been as important as expected, as seed germination seems as a response with a simple output but requiring multiple inputs. Seed sense fluxes in many different environmental factors, and appears to have the capacity to analyze them in a holistic way and to determine whether they germinate or not.

It seems that the presence of an adequate quantity of water and electron flow should take numerous metabolic switches, but in a very precise and complex way, both in space, time, and location. And so much so, that despite the best efforts of researchers in the area, the process of maturation and germination of the seed has not been deciphered yet.

6. Perspective

Despite significant progress in seed biology, basic questions still need to be answered. Further, it is important to translate existing knowledge into agricultural outputs [7].

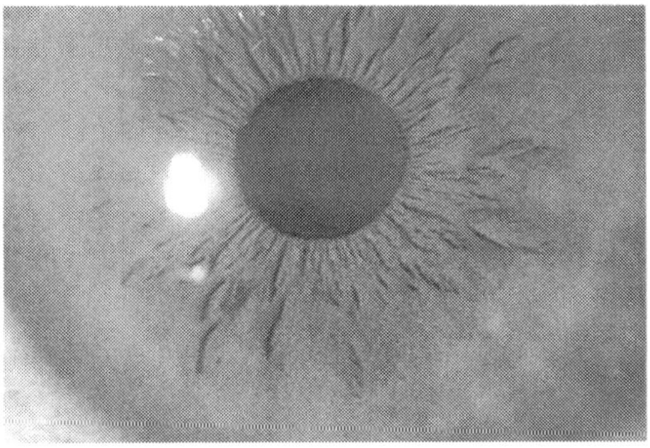

Figure 20. Eye iris. The surface and color of the iris has a remarkable resemblance with the surface and color of several seeds. The color of the eye iris is also due to melanin content.

The insistence of nature in melanin in all forms of life (**Figure 20**) is a message that is very important and essential from which we must learn as much as possible. Let us remember that nature only insists on things that are important and relevant.

And we can say that with the discovery of the intrinsic capacity of melanin to transform the visible and invisible light to chemical energy through the water molecule dissociation, the best is yet to come.

Acknowledgements

The Center for Studies of the Human Photosynthesis® funded this work.

Author details

Arturo Solís Herrera

Address all correspondence to: comagua2000@yahoo.com

Human Photosynthesis® Research Center, Bosques del Prado Norte, Aguascalientes, México

References

[1] Brenner, M., Hearing, V.J. The protective role of melanin against UV damage in human skin. Photochem Photobiol 2008; 84(3): 539–549. Doi: 10.1111/j.1751-1079.2007.00226.x.

[2] Bustamante, J., Bredeston, L., Malanga, G., Mordoh, J. Role of melanin as a scavenger of active oxygen species. Pigment Cell Res 1993; 6: 348–353 [PubMed: 8302773].

[3] Hvam, E., Dahle, J. Melanin synthesis may sensitize melanocytes to oxidative damage by ultraviolet A radiation and protect melanocytes from direct DNA damage by ultraviolet B radiation. Pigment Cell Res 2004; 17: 549–550 [PubMed: 1535783].

[4] Solís-Herrera, A., Arias-Esparza, M.C., Solís-Arias, R.I., et al. The unexpected capacity of melanin to dissociate the water molecule fills the gap between the life before and after the ATP. Biomed Res 2010; 21(2): 224–227.

[5] Solís-Herrera, A. Photoelectrochemical method of separating water into hydrogen and oxygen, using melanins, analogues or derivatives, thereof; as central electrolyzing element. Patent No.: US 8455145 B2. June 4, 2013.

[6] Nambara, E., Nonogaki, H. Seed biology in the 21st century: perspectives and new directions. Plant Cell Physiol 2012; 53(1): 1–4. Doi: 10.1093/pcp/pcr184

[7] Martínez-Andújar, C., Martin, R.C., Nonogaki, H. Seed traits and genes important for translational biology—highlights from recent discoveries. Plant Cell Physiol 2012; 53: 5–15.

Chapter 7

Coumarin-Based Heteroaromatics as Plant Growth Regulators

Fauzia Anjum Chattha, Mehr-un-Nisa,
Munawer Ali Munawer and Samina Kousar

Additional information is available at the end of the chapter

http://dx.doi.org/10.5772/64854

Abstract

Heterocyclic compounds are the largest and most versatile class of organic compounds. Plants produce valuable heterocyclic substances called phytohormones to carry out the growth process. They control the growth or other physiological functions. Both hormones and vitamins are generally referred as plant growth regulators. Aromatic compounds having lactone ring are called coumarin. Coumarin is one of the most important natural substances in plants and is referred to as anti-auxins, as these compounds are considered to play an essential role in plants growth as well as defense.

Keywords: coumarin, inhibitors, phosphorous, heteroarylacetic acids, lactones, glycosides

1. Introduction

Almost all the plants contain some substances which control their growth process and development. These plant growth regulators facilitate the growth processes in a better way to meet the requirements of food supply in general. However, with the increase of food demand and invention of tissue culture in plant science, it was obvious that more and more growth regulators should be designed. A vast majority of growth regulators have been synthesized and tested for their effects on plant growth. In the proceeding paragraphs, the effects of coumarins and derived compounds have been described.

The plant growth regulating activity can be defined as bustle of an active ingredient on the physiological processes of the plant hormones directly responsible for growth reliant on the method to use, development stage on which preferably used and concentration of the active substance used in vitro. The compositions can be prepared by mixing of the active ingredients with some suitable transporters, that is, liquid solvents or solid carriers and optionally surfactants or emulsifiers.

1.1. Coumarins

Coumarins (**1**) are heteroaryl compounds containing lactone ring and are of great medicinal importance. These compounds are supposed to intricate in the various actions comprising plant growth hormones and growth regulators in order to rheostat the respiration, photosynthesis, as well as protection against infection. They have an imperative role in plant biochemistry and physiology, stand-in as antioxidants, enzyme inhibitors and precursors of toxic substance [1].

Coumarins are commonly found in almost every plant family [2]. Plants probably use them as growth inhibitors (anti-auxins) [3, 4] as well as defense compounds mean play an important role in plants' defense system against pests and diseases, including root parasitic nematodes [5]. In some plant families, such as Leguminosae (bean family), Rutaceae (citrus family) [6] and Umbelliferae (Apiaceae) (parsley-fennel family), coumarins are produced and used in larger quantities [7]. The effect of coumarins, umbelliferone (**2**) and xanthotoxins (**3**) is once compared in cucumber, maize and garden pea, and the results are simply thought provoking. Umbelliferone retards root growth less strongly than coumarin and xanthotoxins.

All plants show dissimilar response depending on their Species, size and may be associated with different effects on the process of cellular respiration and enormous changes happening in mitochondria [8].

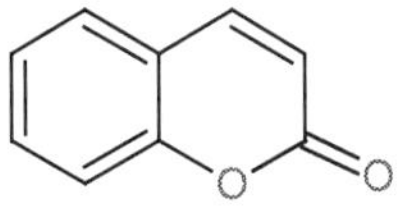

1

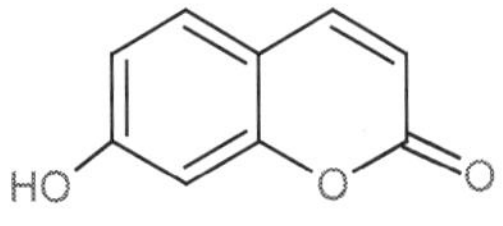

2

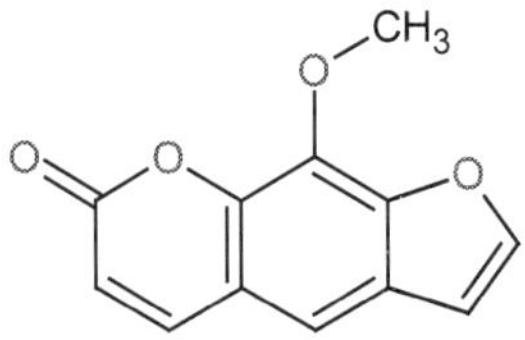

3

1.1.1. *Effects on seed germination*

Abscisic acid (ABA) is a plant growth inhibitor, and indole butyric acid (IBA) is a plant growth activator. At 100 ppm, both ABA and IBA suppress the germination of wheat and sorghum seeds, although this effect is more pronounced in ABA rather than IBA. At 10 ppm, germination percentage is 60–90%, whereas at 1 ppm germination in IBA is >90% in both wheat and sorghum but ABA shows 75–90% germination.

The behavior of coumarin derivatives toward the germination is inhibitory [9, 10]. Different coumarin derivatives show 70–95% germination at 10 and 1 ppm in comparison. 7-Hydroxy-coumarin (**4**) and 7-hydroxy-4-methylcoumarin (**5**) almost completely inhibit the germination in wheat and sorghum seeds at 100 ppm, and percentage germination used to be <50%. All other compounds show 70–95% germination at 10 and 1ppm in comparison to control [11] (**Figure 1**).

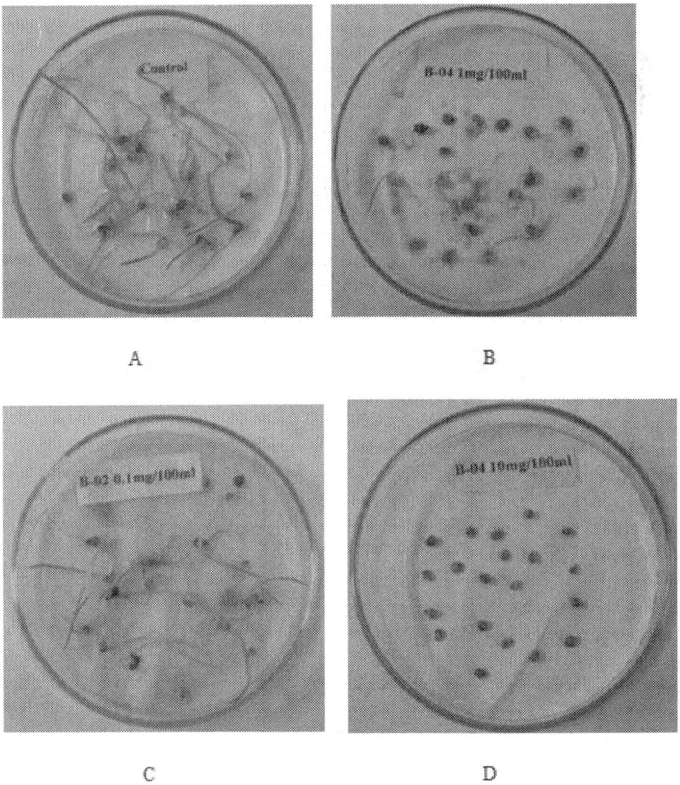

Figure 1. Effect of coumarin derivatives (B = 4, C = 1, D = 5) in comparison with control (A).

Williams et al. [12] have described that hydration and dehydration of radish (*Raphanus sativus* L.) seeds in the presence of coumarin delay the germination and reduce the seedling growth.

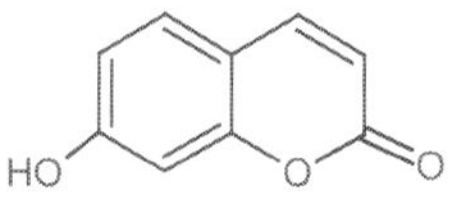

4

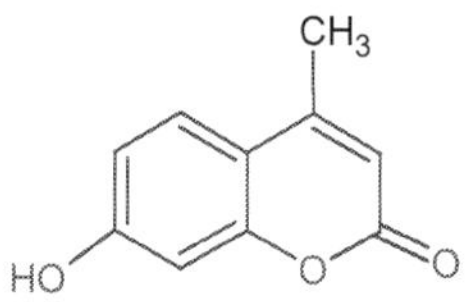

5

Synthesis of coumarin is of much valuable for the plant scientists as coumarin is supposed to play an important role in the growth regulation.

1.1.2. Effects on shoot length

Reduced shoot growth is discerned with 7-methoxycoumarin (**6**), 7-methylcoumarin (**7**) and 7-hydroxy-4-methylcoumarin (**5**) in comparison with that of untreated or control. At 1 ppm concentration, reduced shoot length is perceived by coumarin (**5**) and 7-methylcoumarin (**7**) and 7-hydroxycoumarin (**12**) at 1 and 10 ppm; 7-methoxycoumarin (**6**) and 7-hydroxy-4-methylcoumarin (**5**) at 100 ppm result in 60–85% reduction in shoot growth [13] (**Figure 2**).

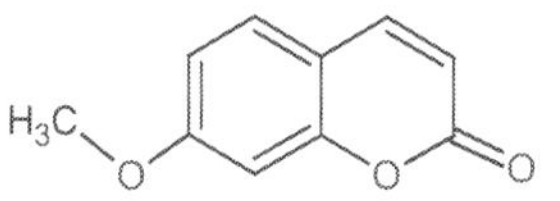

6

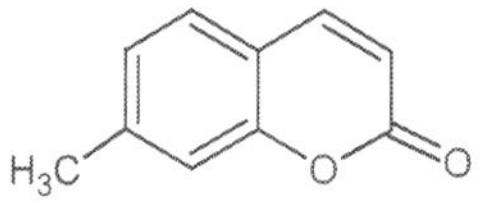

7

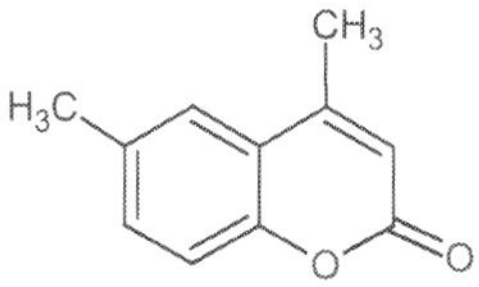

8

Shoot length of sorghum seedlings on the fifth day of germination was completely inhibited by 7-methylcoumarin (**7**), and at 10 ppm, 4,6-dimethylcoumarin (**8**) is the most effective inhibitor of shoot length while other compounds show 20–40% inhibition in shoot growth.

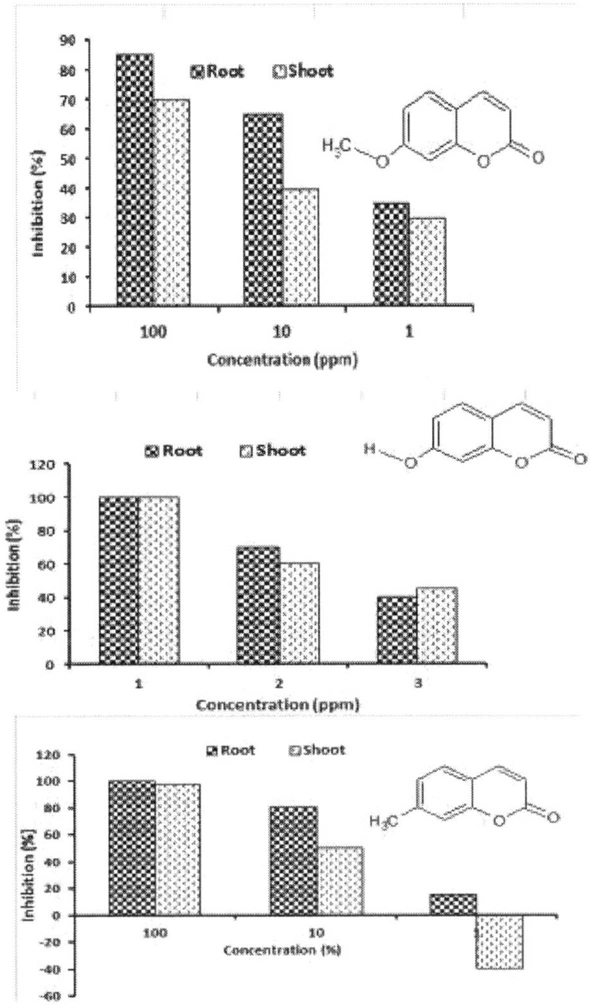

Figure 2. Percentage inhibition of root and shoot growth by 7-methoxycoumarin, 7-hydroxycoumarin and 7-methylcoumarin at various concentrations. Positive bars are growth inhibitors, whereas negative are growth activators.

1.1.3. Effects on root length

Coumarin inhibits the radicle, seminal and nodal root lengths by 50% in solutions of 6, 1 and 0.25 mM [14]. Coumarin decreases the number of lateral roots. The branching density is usually affected more in the seminal than in the radicle roots. The order of sensitivity to coumarin observed to be as: nodal > seminal > radicle roots.

Few reports have been published on the relation between the structure and activity of coumarin and its derivatives. Goodwin and Taves [15] have shown that coumarin is the most powerful root growth inhibitor but some of its derivatives are almost as active as coumarin itself, namely 7,8-dihydroxy coumarin, 7,8-dihydroxy-4-methylcoumarin, 8-methyl coumarin and coumarin-3-carboxylic acid. However, 3-methyl substitution greatly diminishes the inhibitory action on root growth.

Root growth inhibition is about 70-90% by 7-methoxycoumarin at different concentration (1, 10, 100 ppm) and 4,6-dimethylcoumarin and 7-hydroxy-4-methylcoumarin possess the least root growth inhibition activity at 100 ppm. At 10 ppm, 70–90% root length is inhibited by 7-methoxycoumarin and 7-methylcoumarin (**Figure 2**). 6-Hydroxycoumarin, 6-methylcoumarin and 4,6-dimethylcoumarin demonstrate no effect on root length and are seen closer to the control value.

7-Methoxycoumarin is the most effective inhibitors of root growth at 100 ppm compared with that of control. 7-Hydroxy-4-methylcoumarin shows accelerated root growth at 100 ppm. While at 10 ppm, 7-methoxycoumarin, 7-hydroxycoumarin and 7-methylcoumarin inhibit root growth but 6-hydroxycoumarin, 4,6-dimethylcoumarin and 7-methoxy-4-methylcoumarin (**9**) show more growth of seedlings than the control. At 1 ppm, 6-methylcoumarin (**10**), 7-methoxy-4-methylcoumarin and 6-hydroxy-4-methylcoumarin (**11**) are active than that of control.

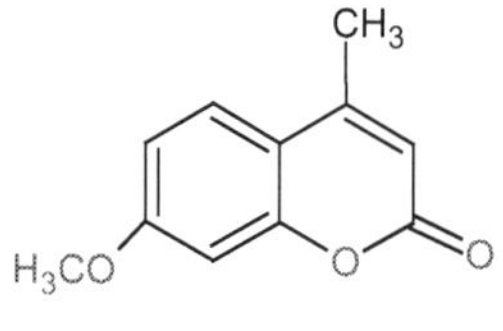

9

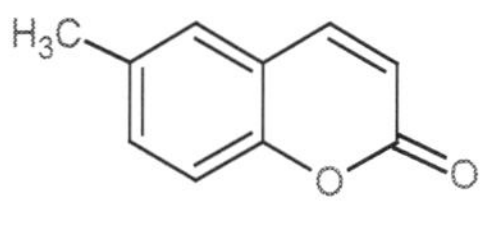

10

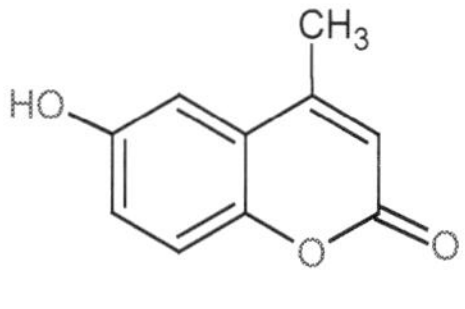

11

1.2. Phosphorous-containing coumarins:

The plant growth regulating activity of phosphorus derivatives of coumarin is of discernible in comparison to parent coumarin as it inhibits the stem growth of pea plants. The elongation of wheat coleoptile segments is also affected by coumarin derivatives. Almost all compounds at concentrations 0.1 and 0.01mM causes stimulation of the pea plant root fresh weight and the stimulating effect reaches 40–49% in ethyl-7-(diethylamino)-2-ethoxy-2*H*-1,2-benzoxaphosphinine-3-carboxylate-2-oxide (compounds 12) and 2-ethoxy-2*H*-1,2-benzoxaphosphinine-3-carboxylate-2-oxide (compounds **13**) while reaches 62% for chloro derivatives (compound **14**).

The influence of the compounds on the growth of cucumber and wheat seedlings is apparent that all phosphorous-containing derivatives inhibit the cucumber root and hypocotyl growth with more than 95% at 1mM, but the effect on wheat roots is weaker and growth reduction is about 80%. Inhibitory activity declines with the decrease of the concentration applied. At concentrations 1 and 0.1 mM, all the compounds exceed the effect of the standard coumarin. Derivatives containing phosphorus in the ring possess strong inhibitory activity at lower concentrations [16].

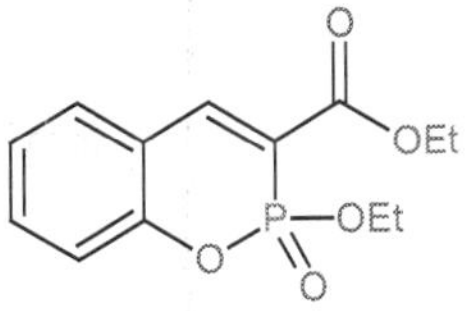

12

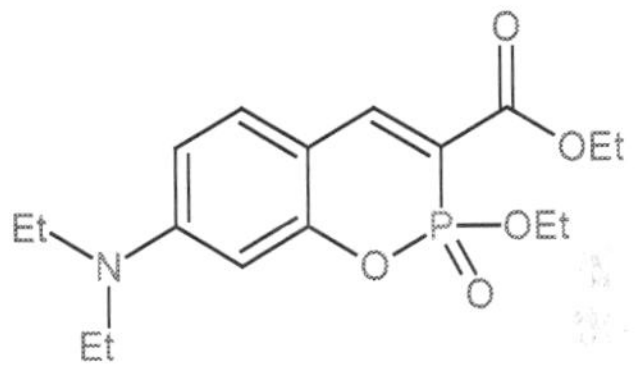

13

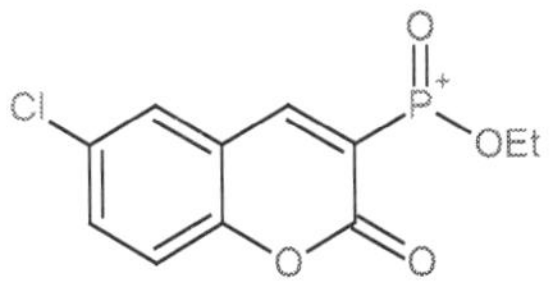

14

1.3. Coumaryl-β-D-glucopyranuronic acid

The effect of coumarin on germination and growth has been studied widely, but very little is known about the natural derivatives of coumarin which occur in plants. It is assumed that in plants they occur in the form of glycosides, and are frequently physiologically inactive even when present in high concentrations [17].

This principle illustrated in an experiment on the effect of various concentrations of o-coumaryl-β-D-glucopyranuronic (CouGN **15**) acid on shoot formation. When o-coumaryl-β-D-glucopyranuronic acid is cleaved by the action of β-glucuronidase (**16**), o-coumaric acid (**17**) is released. o-Coumaric acid is spontaneously converted to coumarin (**1**). This involves the elimination of an acid group, and thereby an increase of pH to a level at which the activity of the native plant β-glucuronidase is presumed to be reduced (**Figure 3**). o-Coumaryl-β-D-glucopyranuronic acid in the growth medium inhibits shoot regeneration induced by BA_3GN sodium salt but not by BA. Best results are observed by using an o-coumaryl-β-D-glucopyranuronic acid at concentration about 3-4 mM. Several mechanisms are involved in the reduction of shoot formation induced by BA3GN, including an increased pH due to the release of o-coumaric acid, leading to a lower frequency of hydrolysis of BA_3GN and a reduced transport of BA3GN into the cells. It is believed that an increased pH is likely to have been at least partially responsible. It indicates that the selectivity of the positive selection system may be improved by using the introduced β-glucuronidase gene to establish a self-regulating mechanism which can significantly reduce the effect of any background enzyme [18].

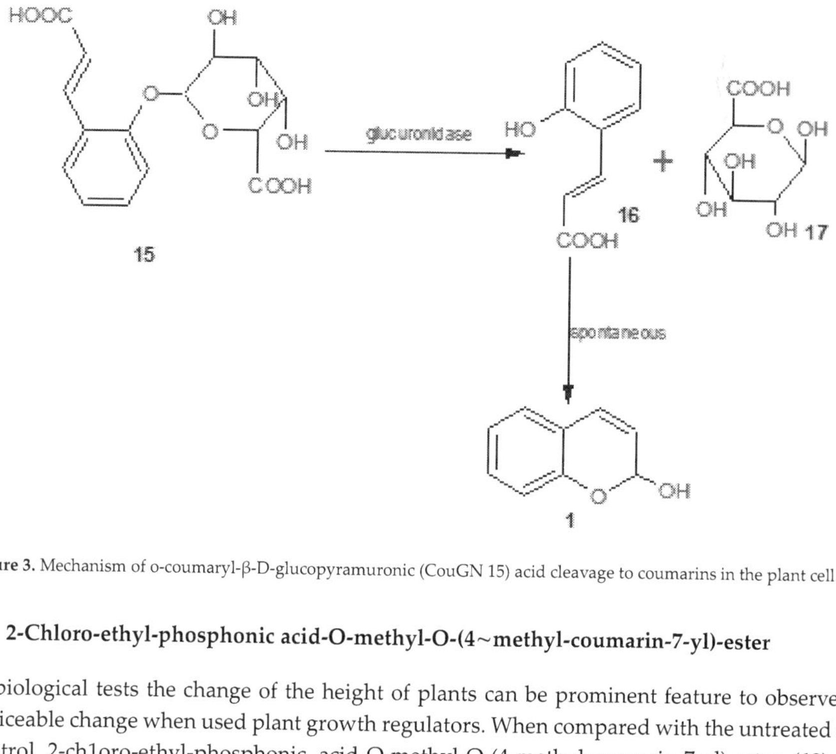

Figure 3. Mechanism of o-coumaryl-β-D-glucopyramuronic (CouGN 15) acid cleavage to coumarins in the plant cell.

1.4. 2-Chloro-ethyl-phosphonic acid-O-methyl-O-(4~methyl-coumarin-7-yl)-ester

In biological tests the change of the height of plants can be prominent feature to observe a noticeable change when used plant growth regulators. When compared with the untreated or control 2-chloro-ethyl-phosphonic acid-O-methyl-O-(4-methylcoumarin-7-yl)-ester (**18**) is supposed to reduce the growth of sunflower at different tested doses. The plant growth is intensively inhibited by 2-chloro-ethyl-phosphonic acid, whereas 7-hydroxy-4-methylcoumarin does not affect the growth of sunflower and double of the same compound slightly inhibited its growth [19].

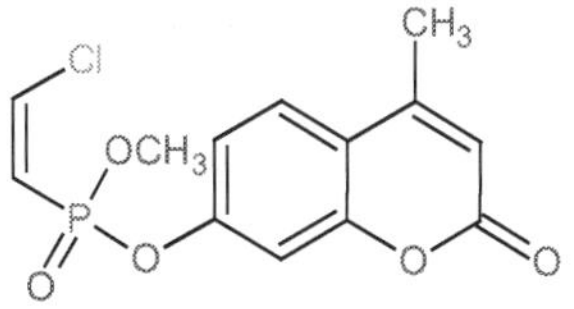

18

A mixture of 2-chloro-ethyl-phosphonic acid and 7-hydroxy-4-methyl, that is 2-chloro-ethyl-phosphonic acid-O-methyl-(4′-methyl-coumarin-7′-yl)-ester, significantly increases the growth of the tomato compared to the untreated or control. It can be seen further that the growth stimulating activity is more intensive than that of 7-hydroxy-4-methyl-coumarin and at higher doses the growth inhibiting activity is lower than that of 2-chloro-ethyl-phosphonic acid.

When using 2-chloro-ethyl-phosphonic acid-O-methyl-O-(4′-methyl-coumarin-7′-yl)-ester in soybeans, all doses reduced the height of soybeans differently from the results observed for tomato. The extent of the reduction of the height is lower than the values measured for 2-chloro-ethyl-phosphonic acid with similar rates. 7-Hydroxy-4-methyl-coumarin stimulated at lower concentration whereas four times of the same compound inhibited the growth. The depressant activity of 2-chloro-ethyl-phosphonic acid is moderated by the 7-hydroxy-4-methyl-coumarin at various concentrations [20].

2. Coumarinacetic acids

Coumarins are naturally occurring compounds. A lot of coumarins have been identified from natural sources, especially green plants. The pharmacological and biochemical properties and therapeutic applications of coumarins depend upon the pattern of substitution [21]. Coumarins have attracted intense interest in recent years because of their diverse pharmacological activities [22–24]. Among coumarin derivatives, Coumarin acetic acids are scarcely reported in the literature.

2.1. Effects of coumarin-3-acetic acids on seed growth

Coumarin-3-acetic acid and its different derivatives were synthesized and tested for their seed germination and plant growth regulating activity. Data show that different derivatives had different effects on seed germination and early growth of plants.

2.2. Effects on seed germination

Coumarin-3-acetic acid (**19**) and 6-aminocoumarin-3-acetic acid (**20**) exhibit 80–95% germination at 100 ppm. Wheat and sorghum seeds possess germination inhibitory activity by the certain coumarin-3-acetic acid derivatives. 5-Hydroxy-4,7-dimethylcoumarin-3-acetic acid

(**21**), 4,6-dimethylcoumarin-3-acetic acid (**22**), 4,7-dimethylcoumarin-3-acetic acid (**23**), 7-methoxy-4-methylcoumarin-3-acetic acid (**24**) and 7-chloro-4-methylcoumarin-3-acetic acid (**25**) at 100 ppm solution show no germination That may be due to reduced water intake that results no imbibition hence seems to possess hindrance in germination [25].

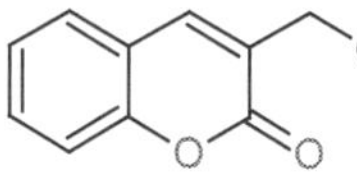

19

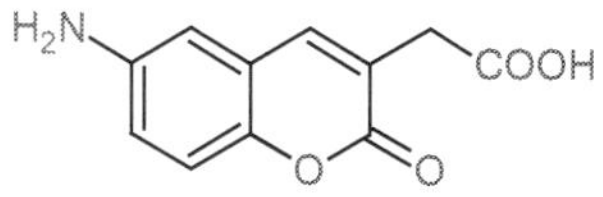

20

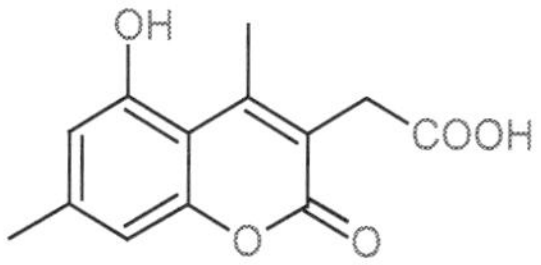

21

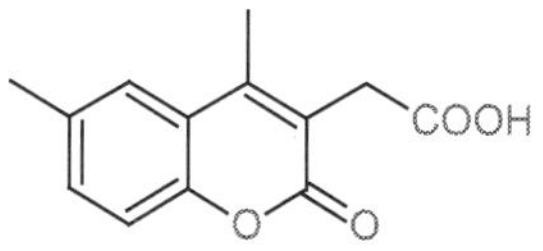

22

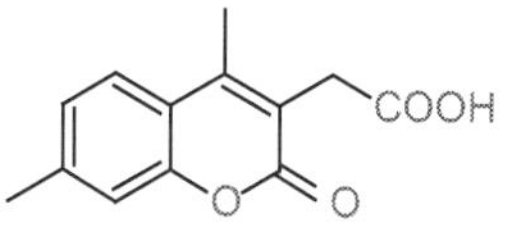

23

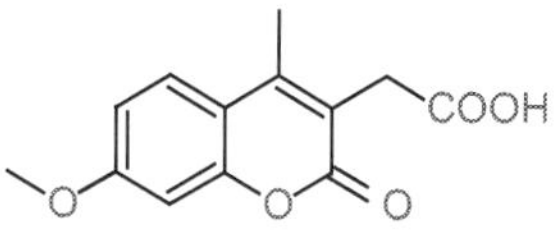

24

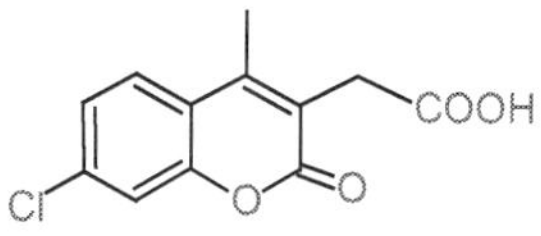

25

2.3. Effects on shoot length

Shoot length (mm) of developing seedlings measured on the fifth day of germination shows inhibition trend. At 100 ppm, 6-nitrocoumarin-3-acetic acid (**26**) completely inhibits wheat seed germination and hence shoot or root growth is usually not observed. Reduced shoot growth is perceived with coumarin-3-acetic acid, and 6-aminocoumarin-3-acetic acid exhibits shoot growth comparable with that of control. At 10 ppm, reduced shoot growth is experiential in 7-hydroxy-4-methylcoumarin-3-acetic acid (**27**) and 7-chloro-4-methylcoumarin-3-acetic acid (**Figure 4**).

http://dx.doi.org/10.5772/64854

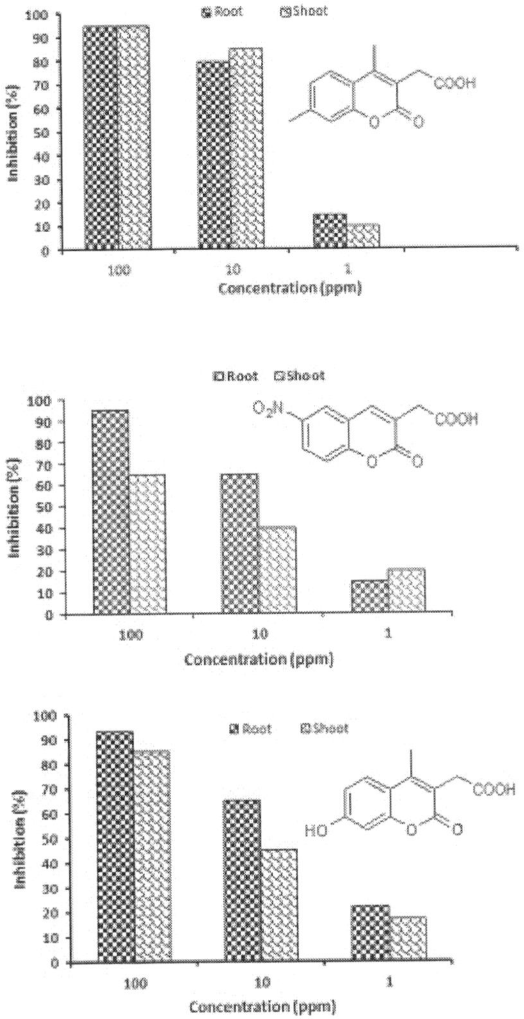

Figure 4. Percentage inhibition of root and shoot growth of seed by action of 4, 6-dimethylcoumarin-3-actic acid **(23)**, 6-nitrocoumarin-3-acetic acid **(26)** and 6-nitrocoumarin-3-acetic acid **(26)** and 6-hydroxy-4-methylcoumarin-3-acetic acid **(27)**.

Shoot lengths of sorghum seedlings on the fifth day of germination are absolutely repressed by **25** and severely inhibited by **24** at 100 ppm. At 10 ppm, 7-hydroxy-4-methylcoumarin-3-acetic acid is the most effective inhibitor of shoot length while other compounds show 20–40% inhibition in shoot growth [25].

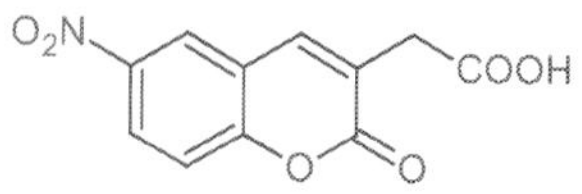

26

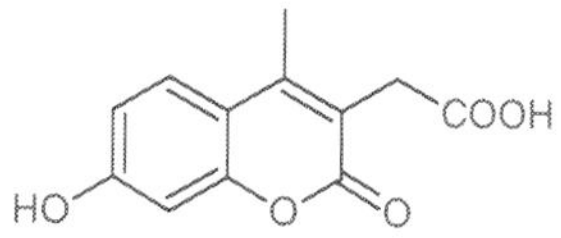

27

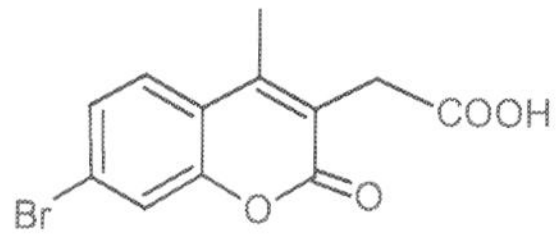

28

2.4. Effects on root length

Changes in root length (mm) monitored on the fifth day of wheat seed germination show inhibition activity. 6-Nitrocoumarin-3-acetic acid completely subdued root growth at 100 ppm. Root growth is impeded between 70 and 90% by coumarin-3-acetic acid and 6-aminocoumarin-3-acetic acid at these concentrations, and 7-bromo-4-methylcoumarin-3-acetic acid (**28**) has the least effect on root growth. At 10 ppm, 70–90% root length remains inhibited by 6-nitrocoumarin-3-acetic acid, 7-hydroxy-4-methylcoumarin-3-acetic acid, 7-methoxy-4-methylcoumarin-3-acetic acid and 7-chloro-4-methylcoumarin-3-acetic acid (**Figure 4**).

Changes in root length (mm) in sorghum seedlings monitored on the fifth day of germination stay constrained by coumarin-3-acetic acid, 6-nitrocoumarin-3-acetic acid and 6-aminocoumarin-3-acetic acid at 100 ppm compared with the control. However, root growth induction is by 20-40% by 7-bromo-4-methylcoumarin-3-acetic acid (**Figure 5**) at 10 ppm [25].

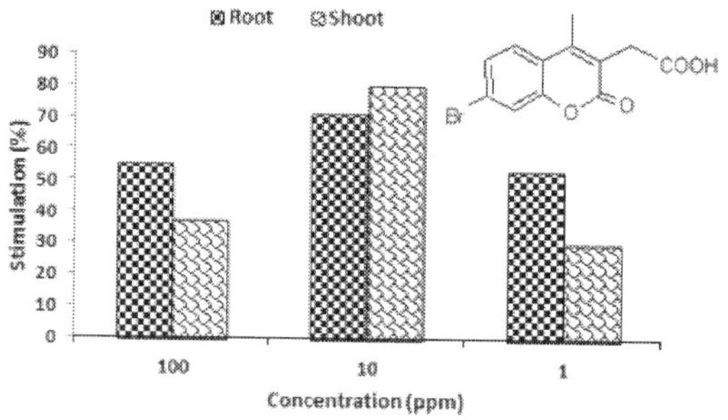

Figure 5. Percentage stimulation of root and shoot growth of seed by the action of 7-bromo-4-methylcoumarin-3-acetic acid (**28**).

http://dx.doi.org/10.5772/64854

3. Percent inhibition/stimulation of growth

Percentage inhibition/stimulation of coumarin-3-acetic acids clarifies the difference of its effect in comparison to control. Most of the compounds possess root growth but less inhibitory effects or somewhat stimulation with respect to the control. In wheat, compounds 6-aminocoumarin-3-acetic acid, 5-hydroxy-4,7-dimethylcoumarin-3-acetic acid, 4,6-dimethylcoumarin-3-acetic acid, 4,7-dimethylcoumarin-3-acetic acid, 7-methoxy-4-methylcoumarin-3-acetic acid and 7-chloro-4-methylcoumarin-3-acetic acid are growth inhibitors. 6-Nitrocoumarin-3-acetic acid, 7-hydroxy-4-methylcoumarin-3-acetic acid, 7-methoxy-4-methylcoumarin-3-acetic acid and 7-chloro-4-methylcoumarin-3-acetic acid at lower concentration (10 and 1 ppm) show very good inhibition of both shoot and root growth.

In sorghum, coumarin-3-acetic acid, 6-nitrocoumarin-3-acetic acid, 6-aminocoumarin-3-acetic acid, 7-hydroxy-4-methylcoumarin-3-acetic acid, 4,7-dimethylcoumarin-3-acetic acid and 7-chloro-4-methylcoumarin-3-acetic acid are clearly substantiated to be growth inhibitors of root, whereas 7-bromo-4-methylcoumarin-3-acetic acid ascertain to be growth stimulator of both root and shoot contrary to its effect in coumarin[25].

Author details

Fauzia Anjum Chattha[1], Mehr-un-Nisa[4*], Munawer Ali Munawer[2] and Samina Kousar[3]

*Address all correspondence to: mehrunnisa20@yahoo.com

1 Govt Islamia P.G. College Cooper road, Lahore, Pakistan

2 University of the Punjab, Lahore, Pakistan

3 University of Engineering and Technology, Lahore, Pakistan

4 University of Education, Lahore, Pakistan

References

[1] Kostova, I. Synthetic and Natural Coumarins as Cytotoxic Agents. Curr. Med. Chem. 2005;5:29-46.

[2] Murray, R. D. H.; Mendz, J. and Brown S. A. The natural coumarin: occurrence. In: John Wiley and sons, editors. Chemistry and Biochemistry. Chichester: John Wiley and sons; 1982.

[3] Pergo, E. M., D. Abrahim, P. C. Soares, K. A. Kern, L. J. D. Silva, E. Voll and E. L. I. Iwamoto. *Bidens pilosa* L. exhibits high sensitivity to coumarin in comparison with three other weed species. J. Chem. Ecol. 2008;34(4):499-507.

[4] T Himann, K. V. and W. D. Bonner, J.R. Inhibition of plant growth by protoanemonin and coumarin, and its prevention by PA. Proc. Natl Acad. Sci. U.S.A. 1949;35:272-276.

[5] Wuyts N, De waele D, Swennen R. Extraction and partial characterization of polyphenol oxidase from banana (Musa acuminate Grande naine) roots. Plant Physiol. Biochem. 2006;44:308-314.

[6] Jiwajinda, S., V. Santisopasri and H. Ohigashi. Coumarin-related compounds as plant growth inhibitors from two rutaceous plants in Thailand. Biosci. Biotechnol. Biochem. 2000;64(2):420-423.

[7] Goren, R. and Tomer, E. Effect of seselin and coumarin on growth, indoleacetic acid oxidase and peroxidase with special reference to cucumber radicles. Plant Physiol. 1971;47:312-316.

[8] Kupidlowska, E.; Kowale, M.; Sulkowski, G. and Zobel, A. M. The effect of coumarins on root elongation and ultrastructure of meristematic cell protoplast. Ann. Bot. 1993;73:525-530.

[9] Knypl, J. S. Action of (2-chloroethyl) trimethylammonium chloride, 2,4- dichlorobenzyltributyl phosphonium chloride, and *N*-dimethylamino-succinamic acid on IAA and coumarin induced growth of sunflower hypocotyl section. Acta Soc. Bot. Pol. 1996;35:611-625.

[10] Knypl, J. S. Growth retardants in relation to germination of seeds. III. The synergistic inhibitory effect of (2-chloroethyl) trimethylammonium chloride and coumarin on germination of kales seeds, and its reversal by kinetin and gibberellic acid. Acta Soc. Bot. Pol. 1967;36:235-250.

[11] Mayer, A. M. And A. Pouakoff-Mayber. Coumarins and their role in growth and germination. In: Ames, editor. Plant Growth Regulation. 4th International Conference on Plant Growth Regulation, Iowa State. Iowa State: Iowa State University Press; 1961. p. 735-749.

[12] Williams R. D.; Peal, L. K.; Bartholomew, P. W. and Williams S. J. Seed hydration-dehydration in an allelochemical (coumarin) alters germination and seedling growth. Allelopathy J. 2005;15(2):4671-4693.

[13] Chattha, F. A., Munawar, M. A.,., Nisa, M., Kousar, S. 2016, .Plant growth regulating activity of coumarins. Plant growth regulating activity of coumarins. Submitted to Bangladesh Journal of Botany

[14] Abenavoli, M. R.; Cacco, G.; Sorgona, A.; Marabottini, R.; Paolacci, A. R.; Ciaffi, M. and Badiani, N. Coumarin differentially affects the morphology of different root types of maize seedlings. J. Chem. Ecol. 2004;30(9):1871-1873.

[15] Goodwin, R.H. and Taves, C. The effect of coumarin derivatives on the growth of Avena roots. Am. J. Bot. 1949;37:224-231.

[16] Alexieva, V., E. Karanov, R. Nikolova and A. Bojilova. 1995. Plant growth regulating activity of some phosphorus derivatives of coumarin. Bulg. J. Plant Physiol. 21(1): 45-51.

[17] Richard A. Jefferson. Assaying chimeric genes in plants: The GUS gene fusion system. 1987. Plant Molecular Biology Reporter, 5(4):387-405.

[18] Finn, T. O., Robert J. W. Method for the selection of genetically transformed cells and compounds for use in the method. EP 0896063 A2. 1999.

[19] Sandor B. V; Judit B. P.; Jézsef F. V.; Andras H. B-G; Elemér T.; Csaba S.V.; J ézsef K. V.; Endre S. A.; Sandor G. S.; Gyiirgy K.; Andras P. T.; ImreCsatlés, H. Etherified 2-hydroxy-ethyl-phosphonic acid derivatives and plant growth regulating agents containing same as active ingredient. US Patents 4,668,274. 1987.

[20] Sandor B. V; Judit B. P.; Jézsef F. V.; Andras H. B-G; Elemér T.; Csaba S.V.; J ézsef K. V.; Endre S. A.; Sandor G. S.; Gyiirgy K.; Andras P. T.; ImreCsatlés, H. Z-Chloro-ethyl phosphonic acid esters and plant growth regulating agents containing same as Active Ingredient. US Patent No. 4,776,874. 1988.

[21] Hoult, J. R. and Paya, M. Pharmacological and biochemical actions of simple coumarins. Gen. Pharmacol. 1996;27(4):713-722.

[22] Lee, H. K.; Oh, S. R.; Kwon, O. K.; Ahn, K.S.; Lee, J.; Kim, J. C.; Min, B. S.; Joung, H. Isolation of coumarins and ferulate from the roots of Angelica purpuraefolia and its antitumor activity. Phytother Res. 2007;21:406-409.

[23] Napolitane, H. B.; Silva, M.; Ellena, J.; Rodrigues, B. D.; Almeida, A. L.; Vieira, P. C.; Oliva, G. and Thiemann, O. H.Aurapten, a coumarin with growth inhibition against Leishmania major promastigotes. Braz. J. Med. Biol. Res. 2004;37(12):1847-52.

[24] Xu, Z. H.; Qin, G. W. and Xu, R. S.A new bicoumarin from *Stellera chamaejasme*. J. Asian Nat. Prod. Res. 2001;3(4):335-340.

[25] Chattha, F. A., Munawar, M. A., Ashraf, M., Nisa, M., Ahmad, S. Plant growth regulating activity of coumarin-3-acetic acids. Allelopathy J. 2015;36(2):225-236.

Chapter 8

Influence of Rootstock on Citrus Tree Growth: Effects on Photosynthesis and Carbohydrate Distribution, Plant Size, Yield, Fruit Quality, and Dwarfing Genotypes

Mary-Rus Martínez-Cuenca,
Amparo Primo-Capella and
Maria Angeles Forner Giner

Additional information is available at the end of the chapter

http://dx.doi.org/10.5772/64825

Abstract

Citrus species are the most widely produced fruit crops in the world, and Spain is one of the leading citrus producers that supply the fresh market. Rootstocks greatly influence variety behaviour as it ensures tolerance to abiotic stress conditions, as well as the provision of minerals and water for the total plant, and consequently impact crop yield and fruit quality. So, rootstock choice is one of the most important decisions a grower makes when establishing commercial citrus orchards. In this chapter, we attempted to provide an overview of the response in terms of plant growth, fruit quality and yield parameters of several citrus cultivar trees grafted onto different commercial rootstocks, plus new hybrids and some dwarfing genotypes, to reduce costs in some cultural practices. In particular, we considered the rootstock influence on scion photosynthetic capacity linked to carbohydrate distribution for plant vegetative and reproductive development.

Keywords: breeding, citrus, drought, dwarfing, Fe chlorosis, flooding, plant growth, rootstock, salinity, yield

1. Introduction

Spain is one of the leading citrus producers that supply the fresh market worldwide. There is a huge variety of cultivars that gives rise to vigorous trees, produces high-quality fruit and allows to extend the commercial period for these fruits from September (earlier clementines, *Citrus reticulata* Blanco) to May (late oranges, *Citrus sinensis* L.). However, several environmental factors could threaten the citrus industry.

The main factors that limit citrus growth include Citrus tristeza virus (CTV) and *Phytophthora* spp., which are present in almost all Spanish citrus orchards. Some abiotic stresses, such as salinity and flooding, also reduce citrus growth in different citrus areas. Moreover, much soil is calcareous and frequently contains over 30% $CaCO_3$ with pH values between 7.5 and 8.5, which causes Fe-deficiency in plants. For these reasons, the trees grown on the rootstocks currently used in Spain face certain problems. Carrizo citrange [*Citrus sinensis* (L.) Osb. × *Poncirus trifoliata* (L.) Raf.], currently the most important rootstock used in Spain, is susceptible to salinity and lime-induced chlorosis. Cleopatra mandarin (*C. reshni* Hort. ex. Tan.) is tolerant to these problems, but tends to grow slowly for the first few years after planting, and induces low yield and fruit size in some varieties. Trees on *C. volkameriana* Ten. & Pasq. are vigorous and bear precociously, but are more susceptible to *Phytophthora*. Thus, attempts have been made to solve abiotic problems through citrus-rootstock breeding programmes worldwide.

Choice of rootstock is among the most important decisions a grower makes, and implications for yield and quality are enormous. Drivers of rootstock adoption are wide-ranging with the most important being tolerance to CTV, *Phytophthora*, nematode and salt, but water-use efficiency and drought tolerance are increasingly becoming important to achieve better performance (**Table 1**). Although the metabolic functions in a grafted plant are divided

	Carrizo citrange	Swingle citrumelo	Cleopatra mandarin	*Citrus volkameriana*	FA 5	FA 13	FA 517	FA 418
Pests and diseases								
Tristeza	R	T	T	T	R	R	R	T
Phytophthora	●●●●	●●●●●	●●●	●	●●●●	●●●	●●●●	●
Nema	S	R	S	S	R	S	R	S
Abiotic stress								
Salinity	●	●●●	●●●●	●●●	●●●●●	●●●●●	●●●●	●●●
Clay soil	●●	●	●●●●	●●●●	●●●	●●	●●●●	●●
Flooding	●●●	●●●●●	●	●●●	●●●●	●●●●		
Horticultural traits								
Tree size	●●●●	●●●●	●●●●	●●●●	●●●●	●●●	●●	●
Yield	●●●	●●●	●●	●●●●	●●●●	●●●●	●●●●	●●
Fruit size	●●●	●●	●●	●●●●	●●●	●●●	●●●	●●●●
Maturity	●●●●	●●	●●●	●●●●	●●●●	●●●●	●●●●	●●●

Symbols key: good/high (●●●●●) to low/poor (●), resistant (R), susceptible (S), tolerant (T).

Table 1. Tolerance average behaviour of main Citrus rootstocks used in Spain to main pests and diseases and abiotic stress factors influencing on plant growth. Effect of citrus rootstock on several horticultural traits related to tree yield and fruit quality.

between the two plant fractions, it is well known that rootstocks greatly influence variety behaviour as it ensures provision of minerals and water for the total plant.

This work overviews the response in terms of plant growth, fruit quality and yield parameters of several citrus cultivars trees grafted onto different commercial rootstocks, plus new hybrids grown under field conditions. The use of some dwarfing genotypes, in which small tree size helps to reduce costs in some cultural practices such as pruning and harvesting, was also considered. This work also reveals the relation between gas exchange parameters, carbohydrate distribution during the annual cycle or hydraulic conductance in roots, shoots and graft union segments as pivotal factors for regulating vegetative and reproductive development. In conclusion, the influence of rootstocks on scions photosynthetic capacity may be a key consideration when determining citrus plant performance in terms of vigour, crop load and fruit characteristics.

2. Factors that limit growth in citrus orchards

2.1. Salinity

Citrus are frequently cultivated in semiarid areas where many soils are either affected by salt or present a high salinisation risk. Soils are considered saline when ECe (electrical conductivity of soil saturated paste extract) is 4 dS m^{-1} or more, which is approximately the equivalent to 40 mM NaCl and generates an osmotic pressure of about 0.2 MPa [1]. Citrus is considered a salt-sensitive crop [2], which suffers physiological disturbances and growth reduction even at low to moderate salinities (**Figure 1**).

Figure 1. Carrizo Citrange leaves affected by salinity stress. (A) Control; (B) salt stressed symptoms (leaf yellowing and tip burnts).

Salinity reduces growth and causes physiological disorders [3, 4], partly induced by adverse water relations due to a reduced soil solution osmotic potential, which alters leaf gas exchange parameters. Salt stress has been shown to decrease water potential (ψ_W), stomatal conductance (gs), transpiration (*E*) and net CO_2 assimilation rates (A_{CO2}) in leaves [5–7]. Excessively high concentrations of saline ions in leaves cause specific toxicities and nutrient imbalance [6, 8].

Major differences in salt stress tolerance have been found between species and family members. Uptake and/or transport of saline ions to the scion is controlled by the rootstock, which chiefly determines chloride (Cl^-) and sodium (Na^+) accumulation in leaves [6, 9–11]. Since the main ion that causes damage is Cl^- [12, 13], the salt tolerance of some citrus rootstocks is usually established by their capacity to exclude Cl^- from leaves [5]. In addition, tolerance to other salt-related stresses, like B toxicity, is likely based on the down-regulation of the main B transport channels in the root, *NIP5* and *PIP1* genes, the capacity to hold B in an insoluble form in the leaves mainly allocated in cell walls, and compartmentalisation of toxic B from the cytoplasm inside the vacuole due to the up-regulation of aquaporin TIP5 [14–16].

2.2. Fe deficiency

It is estimated that between 20% and 50% of the fruit trees grown in the Mediterranean basin suffer from iron (Fe) deficiency [17]. The most prevalent cause of Fe deficiency in this region is the presence of high levels of carbonate ions in calcareous soils, which lead to a high pH, low Fe availability and the condition known as lime-induced chlorosis [18]. The citrus trees planted in these soils often show signs of severe Fe deficiency or Fe chlorosis because of low Fe availability. Iron deficiency affects the biochemistry, morphology and physiology of the whole plant because Fe is an important cofactor of many enzymes, including those involved in the biosynthetic pathway of chlorophylls, which turns into impaired plant growth [19, 20] (**Figure 2**).

Figure 2. Citrus tree affected by iron deficiency at field conditions. (A) No symptoms; (B) Fe-chlorotic symptoms (yellowing) and defoliation branches.

Under Fe-deficient conditions, citrus, like other dicots, have developed the Strategy I mechanism to increase Fe uptake capacity in the root system, which includes increased rhizosphere acidification and Fe^{3+} reduction through proton-ATPases (H^{+}-ATPase) and ferric chelate reductase (FC-R) enzymes, respectively, and stimulation of Fe^{2+} transport across root cell membranes mediated by a specific iron-regulated transporter, IRT [21]. In the xylem, Fe^{3+} is transported to leaves chelated by low-molecular-weight organic acids, mainly citrate and malate; which accumulate in leaves, xylem and roots in response to Fe deficiency [22].

Similar to other disorders, citrus-growing success under iron chlorosis conditions, such as calcareous soils, depends on the availability of suitable rootstocks that are tolerant to low Fe. Trifoliate orange [*Poncirus trifoliata* (L.) Raf.], sweet orange [*C. sinensis* (L.) Osb.] and Carrizo citrange are all susceptible to lime-induced chlorosis, whereas sour orange (*C. aurantium* L.), Cleopatra mandarin and FA 5 (*C. reshni* Hort. ex Tan × *P. trifoliata*) are more lime-tolerant [18, 21, 23, 24].

2.3. Flooding

Soil flooding has been widely reported to affect large areas of the world, and generally in relation to poor soil drainage combined with excessive rainfall or irrigation. One major constraint that stems from excess water is the progressive reduction in both the soil O_2 concentration and redox potential, which leads to the formation of reduced compounds of either chemical or biochemical origin. Accordingly, flooding effects on plants are related mainly to declining aerobic root respiration that impairs ATP synthesis and which, in turn, disrupts plant metabolism and induces a variety of physiological disturbances that alter plant growth [25], including reductions in water flux from roots, hormonal imbalances, altered carbohydrate distribution, deficient nutrient uptake, early leaf senescence and injury in organs, which sometimes precede plant death (**Figure 3**) [26].

Figure 3. Citrus field affected by flooding stress.

Although the response is variable among species and cultivars, citrus is considered a flooding-sensitive crop and responds to waterlogging by restricting stomatal conductance to prevent water loss [27, 28], likely through hormone-regulation and abscisic acid accumulation in leaves [29]. Under these conditions, net CO_2 assimilation by leaves diminishes [27], which leads to altered carbohydrate distribution [30] and oxidative damage to cells due to excess reactive oxygen species generation [31]. During prolonged soil flooding periods, reduced root hydraulic conductance [28, 32] impairs water uptake, which causes leaf wilting, chlorosis and reduced plant growth [29]. As a result of the root physiology dysfunction, flooding also alters nutrient uptake and, therefore, endogenous concentrations of macro- and microelements can be modified. It has been reported that flooding alters nitrogen (N) pools and their partitioning in citrus as a result of reduced uptake and transport [30]. Moreover, in anaerobic soils, N may be lost through denitrification processes [33] which occur because NO_3^- is the first electron acceptor to be reduced following O_2 depletion [34]. Additionally, waterlogging also prevents potassium (K) uptake and, therefore, lowers K concentrations in leaves [27, 35], whereas it helps the uptake of other elements by roots, such as copper and manganese [35]. Fe nutrition is seriously damaging as anoxia conditions promote reduced Fe^{3+} to Fe^{2+} as a result of a lower soil redox potential [25]. Moreover, Fe uptake and plant growth are diminished through the inactivation of the activity of enzymes proton-ATPase and ferric chelate-reductase [36]. Uncontrolled excess Fe^{2+} uptake in acidic soils leads to very high Fe concentrations in plant tissues and inhibition of root growth due to free radical generation [37].

2.4. Water deficit

Plants undergo water stress either when the water supply to their roots is limited or the transpiration rate becomes intense, primarily caused by water deficit such as a drought or high soil salinity. Every year, water stress on arable plants in different parts of the world disrupts agriculture. Hence, the ability to withstand such stress is of immense economic importance (**Figure 4**).

Plants attempt to adapt to stress conditions with an array of biochemical and physiological interventions. Rootstocks present genetically determined characteristics that affect plant water relations, which include root system distribution, water and nutrient absorption efficiency, anatomy of vascular elements and carbohydrate availability [38–40]. The ability of rootstocks to supply water and nutrients to plants, through differences in root hydraulic conductance, could be the main factor to influence fruit development in citrus trees as it likely determines the strength of the grafted variety and its tolerance to water stress [41, 42]. A reduction in root hydraulic conductance under water deficit conditions is associated with either substantial anatomical modifications, such as the development of Casparian bands and suberin lamellae in the exodermis and endodermis [43] or diminished aquaporin activity [28, 44]. Other mechanisms to withstand water stress are higher root-shoot ratios, fewer and smaller leaves, concentrated solutes and carbohydrates, or increased activity of oxidative stress enzymes in leaf cells [45]. As a result of water stress, citrus reduces stomatal conductance (gs), transpiration rate (E) and A_{CO2} [27, 46].

Figure 4. Citrus tree seriously affected by water deficit stress.

3. Photosynthesis and carbohydrate distribution

Biomass differences between trees are likely related to plants' increased ability to assimilate CO_2 during photosynthesis [47]. Trees on rootstocks that enhance photosynthetic capacity grow more vigorously than on other genotypes [48]. This is reflected on photosynthetic product distribution through the differences induced in scion:stock dry matter ratios, which implies changes in source-sink relationships. A high leaf mass with enhanced A_{CO2} results in the translocation of increased C-compounds (mainly sucrose and starch) from shoots to roots, as ^{13}C labelling experiments have demonstrated [48].

Distribution of photoassimilates comprises a tight competition between vegetative organs and developing fruits. Since the number of flowers produced by citrus species is much larger than the number of fruits harvested, it has been concluded that sucrose supply plays a major role in the regulatory mechanism for citrus fruitlet abscission, mainly at the high intensity time of 'june drop' [49]. This phenomenon is strongly dependent on carbohydrate availability and source-sink imbalances. Consequently, photosynthetic activity appears fundamental to supply the high carbohydrate requirements during fruit set since a drop in A_{CO2} results in lower sugar production and increased fruitlet abscission, as reported recently [48] in 'Navel' orange. Moreover, carbohydrate content in leaves, and consequently source:sink imbalance, strongly regulate A_{CO2}. Hence, sugar accumulation in citrus leaves is the signal that regulates the feedback mechanism to stimulate photosynthesis [50]. The starch concentration in root bark is related with the sink strength of fruits. A higher starch concentration in the root bark of fruits in phase I (active cell division and slow growth) may induce low fruit set [48]. In phase II, the starch concentration in root bark tissue lowers as a result of the high carbohydrate demand for fruit growth. Slow fruit growth in phase III leads to new sugar accumulation in roots.

Finally, other factors that may determine photosynthetic differences between genotypes are related to morphological, such as stomatal density, or biochemical facts, such as ribulose-1,5-biphospate carboxylase activity. The hydraulic conductivity of rootstocks has also been correlated with the CO_2 exchange rates of citrus leaves [39, 51].

4. Rootstock citrus breeding programs

The search for new citrus rootstocks that better perform than those currently used is the major aim of the citrus industry in many countries. New diseases, spread of known diseases and citrus culture under different environmental conditions force the demand for new rootstocks. Attempts have been made in several countries to solve these problems through citrus-rootstock breeding programmes.

Several crosses were made in 1951 using *Poncirus trifoliata* as the male parent at the Citrus Research Centre of the University of California [52]. Some works evaluated many of the hybrids obtained for tolerance to CTV (citrus tristeza virus) [53]. Among the studied hybrids, C-13 (*C. depressa* Hay. × *P. trifoliata*) was rated tolerant. Meanwhile, as far as we know, there have been no reports about the performance of C-13. In addition, the reaction of a hybrid of Shekwasha × Swingle trifoliate orange to nematode *Tylenchulus semipenetrans* Cobb, indicated low levels of nematode infestation of roots [54]. While screening citrus hybrids for cold hardiness, a hybrid of Shekwasha × trifoliate orange was reported as having good tolerance levels [55].

In 1974, J.B. Forner began a citrus-rootstock breeding programme using traditional hybridisations at the IVIA (The Valencian Institute of Agrarian Research, Moncada, Valencia, Spain) to obtain new rootstocks tolerant to CTV, salinity and to lime-induced chlorosis and with resistance to *Phytophthora*. To date, more than 500 hybrids have been evaluated to determine their horticultural performance. Of these, four new hybrid rootstocks are now available for better performance in alkaline soils: Forner Alcaide 5 (FA 5) and FA 13 (*C. reshni* Hort. ex Tan × *P. trifoliata*), FA 418 [Troyer citrange (*C. sinensis* × *P. trifoliata*) *C. deliciosa* Ten.] and FA 517 (*C. nobilis Lour* × *P. trifoliata*). These rootstocks have been tested in calcareous soils. The 'Navelina' trees grafted onto FA 5 or FA 13 yielded 40% more than trees on Carrizo citrange, whereas the trees grafted onto FA 5 or FA 13 produced smaller, but similar quality, fruit than those on Carrizo citrange [56, 57].

5. Dwarfing rootstocks

Citrus, like most fruit tree species, are propagated by grafting onto rootstocks that have been selected for their performance under different edaphic conditions or tolerance to diseases. The ability of dwarfing rootstocks to reduce tree vigour has been cited as an important advantage for fruit-crop growers as it allows management cost savings, and even increased yield per unit area in high-density plantations [58]. Despite its interest, the availability of dwarfing rootstocks in citrus has been scarce until recently, and has been restricted almost exclusively to 'Flying Dragon' (*Poncirus trifoliata* L. Raf var. monstrosa). This plant greatly reduces canopy size, increases yield efficiency and produces good fruit quality when used as a rootstock for any citrus cultivar [59]. Despite its resistance to citrus tristeza virus, *Phytophthora* root rot and citrus nematode, it is highly susceptible to iron chlorosis [60], which has likely limited its diffusion in commercial orchards in large citrus cultivated areas.

Fortunately, a few Forner-Alcaide hybrid selections exist which also confer a dwarfing response on scions, in particular FA 517 (**Figure 5**) and FA 418, whose agronomical behaviour has been tested under field conditions [61–63]. Both rootstocks show lower canopy volumes, but higher yield efficiency when compared with Carrizo citrange, the most extended citrus rootstock in Spain [63]. Moreover, they produce good fruit quality and optimal response when cultured in alkaline soils, one of the main factors that limits crops in Spanish soils [61–63].

Figure 5. Influence of Citrus rootstock on tree size response at field conditions. (A) Tree on Forner-Alcaide 517 (dwarfing behaviour) and (B) Forner-Alcaide 5 (normal size).

Growth reduction induced by dwarfing rootstocks has been associated with lower leaf and stem water potentials in the scions grafted onto them compared with those grafted onto vigorous rootstocks, probably due to high hydraulic conductivity resistance, which may cause water deficit in leaves during periods of high evaporative demand and stomata closure [64]. Consequently, dwarfing rootstocks are poorly able to transport water from soil to stems [65]. Although the resistance of bud union to water transport and xylem anatomical characteristics, in particular the number and diameter of vessels, may limit plant growth, carbohydrate distribution is also an important constraint involved in tree response [63, 64]. So, the reduced translocation of photoassimilates from leaves to roots limits root development and also contributes to the greater availability of these compounds in the scion, which results in increased carbon transport towards fruits. This explains the high yield efficiency and good fruit quality that these rootstocks exert on scions.

6. Rootstock effects on

6.1. Tree vegetative growth

It is widely assumed that rootstocks greatly affect tree size [66–69], which has been noted in some citrus rootstocks (**Table 1**), such as Flying dragon [64, 70]. This effect is of much interest in citrus breeding works as it cuts yield costs. However, growth evaluations in adult trees are complicated by handling difficulties. Some tree morphology-based parameters allow estimates of plant development under field conditions to be made without destroying plant material. Thus, the trunk cross-sectional area (TCSA) is usually considered to be highly correlated with tree weight and canopy volume [71, 72]. A study carried out in Lane Late orange grafted onto different rootstocks has reported that trees on *C. macrophylla* W. had a smaller TCSA and trunk diameter than Gou Tou Chen (Citrus hybrid of *C. aurantium*) and Cleopatra mandarin trees [73]. Similar TCSA values have been obtained on *C. volkameriana* with 'Clementine' mandarin [66].

Another good parameter to evaluate plant development and relative growth between both tree fractions is the scion/stock ratio. It corresponds to the ratio of the circumference from the scion to that of the rootstock, reflects the difference in the growth rates of each tree fraction, and is used as an indicator of scion/rootstock affinity [74]. The closer this ratio is to a value of one, the better affinity between scion and rootstock is observed, and therefore, less interference in tree growth. In oranges, *C. volkameriana* presents a good scion/rootstock affinity with Lane Late scions (0.94), but lowers to 0.88 with *C. macrophylla* and Cleopatra mandarin [68, 73]. In lemons (*Citrus limon* Burn. F.), the best scion/ratio corresponds to combinations with *C. macrophylla.* Its good agronomical behaviour has allowed these rootstocks to become the most widely used for lemon crops in Spain [67, 75].

Nevertheless, the normal cultural practice in Spain is to form scaffold branches next to the bud union, but this makes TCSA measurements difficult. Under these conditions, canopy volume proves to be a better parameter to evaluate tree size. 'Navelate' trees grafted onto *C. volkameriana* and Cleopatra mandarin presents a larger canopy volume than on C-13 [76]. This tendency has also been observed in 'Navelina' orange trees grafted onto *C. volkameriana* rootstocks [57]. Some new rootstocks obtained in the breeding programme carried out at the IVIA, in particular rootstocks FA 5 and FA 13, presented an intermediate size [57], while smaller trees were grated onto FA 418 [77]. In mandarine scions, *C. volkameriana* also conferred the largest size to 'Clausellina' trees compared with Carrizo citrange [78], the most commonly used rootstock in Spain. Once again, mandarin trees on C-13 had the smallest canopy volume and shortest tree height [78].

6.2. Yield and yield efficiency

In general terms, citrus trees yield their first crop 2–3 years after planting and these plants reach full production by year 5 or 6. However, Cleopatra mandarin produces a very reduced number of fruits, which is typical of its slow growth tendency in the first few years after planting, and trees grafted onto this genotype do not reach full production until year 8 [76]. Another exception is FA 418, which is considered to anticipate bear fruit [77].

Table 1 also lists the yield per tree associated with the main Citrus rootstocks. In lemons, *C. macrophylla* generates high crops in most lemon varieties, while other rootstocks like Carrizo citrange or Cleopatra mandarin induce low yields [67, 75, 79], which is likely linked to their low TCSA [72]. Conversely, good yields were observed in 'Eureka' lemon on Cleopatra mandarin rootstocks [80]. Trees on *C. macrophylla* rootstock produced the highest cumulative yield in 'Lane Late' oranges, and showed no significant differences with trees on Cleopatra mandarin. In contrast, a study into 'Marisol' Clementine, reported that Cleopatra mandarin was the least productive rootstock [68]. However, 'W. Navel' orange trees budded onto Carrizo citrange produced the highest fruit yield, while the lowest corresponded to Cleopatra mandarin [81]. In the other hand, fruit yield of some mandarin trees as 'Fallglo' and 'Sunburst' were not affected by rootstocks [82].

Despite the small tree size with the C-13 genotype, its high yield allows this genotype to present good yield efficiency on a canopy volume basis, followed by Carrizo citrange and *C. volkameriana* [8]. At the far end, the low yield of trees on Cleopatra mandarin means very poor yield efficiency. In 'Lane late' trees, the best yield efficiency corresponded to the trees on *C. macrophylla*, and the lowest yield-efficient trees were those grafted onto Gou Tou, Cleopatra mandarin and FA 418, while *C. volkameriana* offered an intermediate yield efficiency [73, 77]. Similar results have been reported on 'Navelina' orange, who found that the trees on *C. volkameriana* had similar yield efficiency, but lower yield efficiency on Cleopatra mandarin [57]. In line with this, studies in 'Marisol' Clementine and found that all the studied rootstocks achieved similar yield efficiency [68]. These results agree with others on 'Shamouti' orange, 'Nova' mandarin and 'Clementine' mandarin, in Cyprus [66, 83]. Interestingly, the low vigour and high yield efficiency of the trees grafted onto FA 517 and FA 418 indicated that these rootstocks were suitable for high-density plantings to compensate for the reduced productivity of individual trees [63]. Other studies also observed good yield efficiency and low TCSA values in 'Eureka' and 'Lisbon' trees on *C. macrophylla* rootstock [75].

6.3. Fruit drop

Fruit drop is a major disorder that comprises fruit yield in citrus orchards. So, growers apply 2,4-DP (2-ethylhexyl ester or dichlorprop-*p*) to reduce this problem. However, this practice increases cultivation costs. In addition, the excess of production of some cultivars prolongs the harvest period and a significant part of the crop is lost. For these reasons, an excellent trait for rootstocks is to retain ripened fruit. The tendency of fruits to drop increases with plant age, which is strongly regulated by the influence of rootstocks on scions. Thus, C-13 ranks the highest for fruit drop, with values from 51% to 85% when used as a rootstock for 'Navelate' orange, but with values from 40% to 65% when grafted onto Carrizo citrange, Cleopatra mandarin and *C. volkameriana* [76]. In 'Lane Late' orange, Cleopatra mandarin and new hybrids FA 030230, FA 020321, FA 418 and FA 030212 showed low pre-harvest drop, while Carrizo citrange, FA 030127 and FA 13 obtained a fruit drop value above 36% [77].

6.4. Alternate bearing index (ABI)

The ABI analysis, which reflects the difference in yield between two consecutive harvests, shows important new information since it normally differs among rootstocks. Some authors have described that there are some rootstocks which increase the alternate bearing of yields [77], and that this effect is extremely harmful in the commercial varieties that already present this defect. This is the case of the 'Verna' lemon variety, which develops a very alternate behaviour [84]. Some authors have indicated that *C. macrophylla* [73] is one of the rootstocks that presents the most uniform productivity (less than 16% ABI). Good ABI values were also obtained for 'Lane late' orange when grafted onto rootstocks Carrizo citrange, FA 5, FA 030131 and FA 020324 [73]. FA 418 also has a poor alternate effect on Navel orange trees [63]. However, the fruits on Cleopatra mandarin and *C. volkameriana* present a relatively low-moderate ABI values (between 23% and 35%), but exhibit more than 50% alternate bearing on Gou Tou [77]. 'Nova' and 'Clementine' mandarin trees within a wide range of most known rootstocks displayed relatively high ABI values [66]. In contrast, some studies into 'Shamouti' orange and 'Fallglo' and 'Sunburst' mandarin have suggested that alternate bearing is not rootstock-dependent [82, 83].

6.5. Fruit quality variables

In fruits destined to be consumed as fresh fruit, fruit size, juice content and the TSS/TA ratio are most important. Rootstocks have been reported to significantly affect both the external (size, rind thickness, peel colour, etc.) and internal (juice content and colour, pH, total soluble solids, etc.) quality variables of fruit. The influence of several Citrus rootstocks on two of the main quality factors (fruit size and maturity) is listed in **Table 1**.

6.5.1. Fruit size

Rootstocks do not apparently affect fruit size in the first harvest years of citrus crops, but do alter fruit size as the tree age increases. After the ninth harvest, the trees on *C. volkameriana*, C-35 citrange (*C. sinensis* cv. Ruby × *P. trifoliata* cv. Webber Fawcett.) and Carrizo citrange produced the heaviest and largest fruits on mandarin and orange fruits [57, 73, 77, 78]. High values have also been obtained in fruits on some hybrid selections, such as FA 030123 and FA 030142. In contrast, the oranges on Gou Tou and Cleopatra mandarin yielded light and small fruits [73], and the same occurred with the FA 13 hybrid [57]. Forner-Giner et al. [8] have also reported small and light fruits on the trees on Cleopatra mandarin and FA 020326 hybrid selection. In contrast, [68] have reported 'Marisol' clementine and [85] on 'Shamouti' oranges that the trees on sour orange, Carrizo citrange and Swingle citrumelo [*C. paradisi* Macf. × *P. trifoliata*] produced similar fruits in both weight and size terms. Meanwhile, Tuzcu et al. [81] have indicated that the fruit weight of 'W. Navel' orange on sour orange was similar to that on Carrizo citrange and Cleopatra mandarin. It is noteworthy that FA 418 has been reported to maintain good fruit growth in orange varieties despite its dwarfing behaviour [63]. In lemons, the trees on *C. macrophylla, C. volkameriana* and other less known rootstocks in Spain (*C. sulcata, C. taiwanica* Tan. Shim. and *C. ampullacea*) have been found to generate large-sized fruits [72, 79]. In contrast, reduced size in lemon has been observed in the fruits of the Cyprus local variety

'Lapitkiotiki' on Cleopatra mandarin, sour orange and Morton citrange [*C. sinensis* cv. Washington navel × *P. trifoliata* (L.) Raf.] rootstocks [67]. *Citrus amblycarpa* Ochse and Cleopatra mandarin rootstocks have also induced a small fruit diameter in 'Eureka' lemon [79].

6.5.2. Rind thickness

The rind thickness analysis is a good parameter to estimate fruit quality as it is inversely correlated with the amount of juice [77]. Rind thickness extremes are not desirable as thick rind is normally related with low juice content, while thin rinds are prone to splitting and are sensitive to peel disorders, which can occur during storage. This parameter is also influenced by rootstock. Thus, the 'Lane Late' fruits from *C. volkameriana* and *C. macrophylla* have thick rinds, which are thinner when grafted onto Gou Tou and Cleopatra mandarin [73]. Although larger, the fruits of the trees on C. *volkameriana* present thick rinds, while C-13 peels are very thin [8]. In oranges, the FA 418 rootstock also confers thick peel to fruit (>4.8 mm), but this parameter drops to near 4.0 mm in the fruits on Cleopatra mandarin, FA 13, Carrizo citrange and some hybrid selections, such as FA 030212, FA 030127 and FA 030113 [8, 57]. In mandarins, once again *C. volkameriana,* and also Carrizo citrange, present the thickest rind (around 2.3 mm), while the C-13 rootstock induces thin peels, around 2.0 mm [78].

6.5.3. Colour index (CI)

Colour is considered one of the most important external factors of fruit quality as fruit appearance greatly influences consumer acceptance. A coloured fruit on the tree is always ripe, so the risk of selecting immature fruit due to colour is highly improbable, unless they are artificially degreened. A non-destructive method exists that can be applied in the field and in industry to accurately show the apparent degree of fruit maturation in temperate countries [86]. According to [73], oranges with the best external colour are produced on *C. macrophylla* and *C. volkameriana* (CI ~ 1.82), and the worst are yielded on Gou Tou (1.16). Forner-Giner et al. [57] also recorded low CI in 'Navelina' oranges on Cleopatra mandarin. In [61], 'Lane late' fruits with the best external colour were produced on Cleopatra mandarin and FA 020324 hybrid rootstocks (CI > 1.30), which showed the most intense orange-coloured skin due to a higher a^* parameter. However, a high L^* parameter in the fruits on FA 030131 and 030127 lowered CI to values under 0.45. Carrizo citrange also resulted in attractive orange fruits with a higher CI than 1. Similar values of L^*, a^* and b^* have been obtained in other studies [87, 73]. The fruit colour index of fruits grafted onto FA 418 was also lower than on other rootstocks [63].

6.5.4. Juice content and colour

In general, the larger the fruit and the thicker the rind are, the lower the juice content is. This applies to the higher juice content of 'Marsh' grapefruits (*Citrus paradise* Macf.) on sour orange than on *C. amblycarpa* or Cleopatra mandarin [88]. In orange, the fruits of *C. volkameriana* and *C. macrophylla* present low juice content [73]. Accordingly, García-Sanchez et al. [10] found that the fruits of 'Clemenules' mandarin on Carrizo citrange had a higher juice percentage and a lower peel percentage than those on Cleopatra mandarin. Contrarily, statistically significant differences in fruit peel thickness and juice content were not found among rootstocks by these authors: [68] on 'Marisol' clementine, [85] on 'Ortanique' tangor, [81] on 'W. Navel' and [82] in 'Fallgo' and 'Sunburst' mandarins. Forner-Giner et al. [8] also reported a higher juice content in the fruits grafted onto Carrizo citrange, C-13 selection and FA 020326 than on Cleopatra mandarin and *C. volkameriana*. Misra et al. [89] obtained a maximum juice content in the fruits of lemon trees onto trifoliate orange and Cleopatra mandarin, while the lowest content went to *C. taiwanica* [90]. Rootstocks influence juice colour. In [61], the fruits of the trees grafted onto Cleopatra mandarin and Gou Tou were more luminous in colour (a higher L^* parameter), while those grafted onto *C. macrophylla* and *C. volkameriana* produced fruit with the most intensely orange-coloured skin (higher a* parameter). Similar L^*, a^* and b^* results were obtained on 'Hamlin' and 'Earlygold' by Lee and Castle [87].

6.5.5. Total soluble solids (TSS) and total acid (TA) percentages

The flavour and palatability of citrus fruits vary according to relative levels of TSS, and also to the presence or absence of aromatic or bitter juice constituents [91]. Carrizo citrange, FA 030212 and FA 030230 selections induced higher TSS in the fruits of 'Lane late' navel orange than Cleopatra mandarin, FA 13 and FA 030127 [77], while the lowest values were found when analysing the fruits from the Gou Tou rootstock [73]. Interestingly in clementines, Cleopatra mandarin induced a higher TSS in fruits than in orange varieties, as reported in 'Marisol' and 'Clemenules' studies [68, 92]. It is noteworthy that the high yield recorded by *C. macrophylla*, did not significantly affect its TSS compared with other rootstocks.

In 'Lane late' orange, a low total acid (TA) percentage was found on the fruits of *C. volkameriana*, and with no significant differences with *C. macrophylla*, while the highest acidity was induced by Cleopatra mandarin and Gou Tou [73]. Carrizo citrange, FA 418 and FA 030212 also induced high TA contents, while low levels were induced in the fruits of FA 13 [77]. In contrast, some authors have found that the effects of rootstock on fruit juice acidity were non-significant [10, 81, 93]. Regarding organic-acid content, *C. volkameriana* and *C. macrophylla* induced low total acid values on 'Lane late' orange [73]. The major organic acid in 'Lane late' navel orange was citric acid (0.89–1.15%) and differences were found between genotypes. In contrast, malic acid was not apparently affected by rootstock, with values between 0.29% and 0.31%. These results were also found on calamondin and 'Kozan' orange [94, 95]. Interestingly, high ascorbic acid values have been recorded in the 'Lane late' trees grafted onto Gou Tou and Cleopatra mandarin [73].

The ripeness index (RI), which relates the soluble solid content measured in °Brix and the titratable acidity determined as a percentage of citric acid content in fruit juice, is the most widespread method used to estimate the citrus fruit maturity level. In 'Lane late', the fruits on FA 13 obtained a high RI value, which was low on FA 418, FA 030212 and Carrizo citrange. The fruits of the trees on *C. volkameriana* and *C. macrophylla* also showed high RI values, which were lower in Cleopatra mandarin and *C. volkameriana* [73]. However on 'W. Navel' orange, on 'Rhode Red Valencia' orange and on 'Okitsu' Satsuma mandarin, [81, 93, 96] reported that the effects of rootstocks on the RI were not statistically significant.

http://dx.doi.org/10.5772/64825

6.5.6. *Sugar content*

Finally, sugars are the major components of citrus juice soluble solids and sweetness of orange juice is intrinsic to its sugar composition. Sucrose is the main sugar present in orange juice, more than 55%, followed by glucose and fructose [73, 95]. The bibliography strongly relates sugar content and rootstock influence on variety. This effect is very important in orange and mandarin fruits [62, 69], and has also been observed in other studies carried out on lemons [75, 79]. Legua et al. [61] also observed high total sugars content in the juice from *C. macrophylla* and Cleopatra mandarin, and low contents from *C. volkameriana*. Similar results have been reported on 'Kozan' and 'Salustiana' oranges [95, 97]. In contrast, some works have found no appreciate significant differences in the juice contents of 'Comune' Clementine, 'Orlando' tangelo and grapefruit on the rootstocks studied therein [98–100].

7. Conclusions

Citrus growth is dependent on rootstock effect. Plant responses to abiotic stress conditions where rootstock behaviour plays a key role in tree development. Rootstock influence on the scion's photosynthetic capacity linked to carbohydrate distribution during the annual cycle is a determining factor for plant vegetative and reproductive development. Therefore, rootstock strongly regulates the plant growth, yield and fruit quality of the cultivars. Finally, new dwarfing rootstocks, in which small tree size helps to cut the costs in some cultural practices such as pruning or harvesting, confers very promising and interesting physical and chemical properties to scions which strongly supports their use for citrus production.

Author details

Mary-Rus Martínez-Cuenca*, Amparo Primo-Capella and Maria Angeles Forner Giner

*Address all correspondence to: martinez_mru@gva.es

Citriculture and Vegetal Production Centre, The Valencian Institute of Agrarian Research (IVIA), Moncada, Valencia, Spain

References

[1] Munns R, Tester M. Mechanisms of salt tolerance. Annu Rev Plant Biol. 2008;59:651–681.

[2] Maas EV. Salinity and citriculture. In: Tribulato E, Gentile A, Refergiato G, editors. In: Proceedings of the International Society of Citriculture, vol. 3, Acireale, Italy; 1992, pp. 1290–1301.

[3] Gómez-Cadenas A, Iglesias DJ, Arbona V, Colmenero-Flores JM, Primo-Millo E, Talon M. Physiological and molecular responses of citrus to salinity. Recent Res Dev Plant Mol Biol. 2003;1:281–298.

[4] Syvertsen JP, García-Sanchez F. Multiple abiotic stresses occurring with salinity stress in citrus. Environ Exp Bot. 2014;103:128–137.

[5] Bañuls J, Serna MD, Legaz F, Talón M, Primo-Millo E. Growth and gas exchange parameters of citrus plants stressed with different salts. J Plant Physiol. 1997;150:194–199.

[6] García-Sanchez F, Jifon J, Carvajal M, Syvertsen JP. Gas exchange, clorophyll and nutrient contents in relation to Na^+ and Cl^- accumulation in "Sunburst" mandarin grafted on different rootstocks. Plant Sci. 2002;162:705–712.

[7] Perez-Perez JG, Syvertsen JP, Botía P, García-Sanchez F. Leaf water relations and net gas exchange responses of salinized Carrizo citrange seedlings during drought stress and recovery. Ann Bot. 2007;100:335–345.

[8] Forner-Giner MA, Legaz F, Primo-Millo E, Forner JB. Nutritional responses of citrus rootstocks to salinity: performance of the new hybrids, Forner-Alcaide 5 and Forner-Alcaide 13. J Plant Nutr. 2011;34:1–16.

[9] Forner-Giner MA, Primo-Millo E, Forner J. Performance of Forner-Alcaide 5 and Forner-Alcaide 13, hybrids of Cleopatra mandarin × *Poncirus trifoliata*, as salinity-tolerant citrus rootstocks. J Am Pomolog Soc. 2009;63:72–80.

[10] García-Sanchez F, Syvertsen JP, Martinez V, Melgar JC. Salinity tolerance of 'Valencia' orange trees on rootstocks with contrasting salt tolerance is not improved by moderate shade. J Exp Bot. 2006;57:3697–3706.

[11] Martínez-Alcántara B, Martínez-Cuenca MR, Quiñones A, Iglesias DJ, Primo-Millo E, Forner-Giner MA. Comparative expression of candidate genes involved in sodium transport and compartmentation in citrus. Environ Exp Bot. 2015;111:52–62.

[12] Hussain S, Luro F, Costantino G, Ollitrault P, Morillon R. Physiological analysis of salt stress behavior of citrus species and genera: low chloride accumulation is an indicator of salt tolerance. S Afr J Bot. 2012;81:103–112.

[13] Ruiz M, Quiñones A, Martínez-Alcántara B, Aleza P, Morillon R, Navarro L, Primo-Millo E, Martínez-Cuenca MR. Effects of salinity on diploid (2×) and doubled diploid (4×) Citrus macrophylla genotypes. Sci Hortic. 2016;207:33–40.

[14] An J, Liu Y, Yang C, Zhou G, Wei Q, Peng S. Isolation and expression analysis of CiNIP5, a citrus boron transport gene involved in tolerance to boron deficiency. Sci Hortic. 2012;142:149–154.

[15] Martínez-Cuenca MR, Martínez-Alcántara B, Quiñones A, Ruiz M, Iglesias DJ, Primo-Millo E, Forner-Giner MA. Physiological and molecular responses to excess boron in Citrus macrophylla. PLoS ONE. 2015;10(7):e0134372. doi: 10.1371/journal. pone. 0134372.

[16] Ruiz M, Quiñones A, Martínez-Alcántara B, Aleza P, Morillon R, Navarro L, Primo-Millo E, Martínez-Cuenca MR. Tetraploidy enhances boron-excess tolerance in Carrizo citrange (*Citrus sinensis* L. Osb. × *Poncirus trifoliata* L. Raf.). Front Plant Sci. 2016;7:701.

[17] Jaegger B, Goldbach H, Sommer K. Release from lime induced iron chlorosis by CULTAN in fruit trees and its characterisation by analysis. Acta Hortic. 2000;531:107–113.

[18] Pestana M, de Varennes A, Abadía J, Faria EA. Differential tolerance to iron deficiency of citrus rootstocks grown in nutrient solution. Sci Hortic. 2005;104:25–36.

[19] Abadía J, Morales F, Abadía A. Photosystem II efficiency in low chlorophyll, iron-deficient leaves. Plant Soil. 1999;215:183–192.

[20] Molassiotis A, Tanou G, Diamantidis G, Patakas A, Therios I. Effects of 4-month Fe-deficiency exposure on Fe reduction mechanism, photosynthetic gas exchange, chlorophyll fluorescence and antioxidant defence in two peach rootstocks differing in Fe-deficiency tolerance. J Plant Physiol. 2006;163:176–185.

[21] Martínez-Cuenca MR, Forner-Giner MA, Iglesias DJ, Primo-Millo E, Legaz F. Strategy I responses to Fe-deficiency of two *Citrus* rootstocks differing in their tolerance to iron chlorosis. Sci Hortic. 2013;153:56–63.

[22] Martinez-Cuenca MR, Iglesias DJ, Talon M, Abadia J, Lopez-Millan AF, Primo-Millo E, Legaz F. Metabolic responses to iron deficiency in roots of Carrizo citrange [*Citrus sinensis* (L.) Osbeck. × *Poncirus trifoliata* (L.) Raf.]. Tree Physiol. 2013;33:320–329.

[23] Martínez-Cuenca MR, Iglesias DJ, Forner-Giner MA, Primo-Millo E, Legaz F. The effect of sodium bicarbonate on plant performance and iron acquisition system of FA-5 (Forner-Alcaide 5) citrus seedlings. Acta Physiol Plant. 2013;35:2853–2845.

[24] Llosa MJ, Bermejo A, Cano A, Quiñones A, Forner-Giner MA. The citrus rootstocks Cleopatra mandarin, *Poncirus trifoliata*, Forner-Alcaide 5 and Forner-Alcaide 13 vary in susceptibility to iron deficiency chlorosis. J Am Pomolog Soc. 2009;63:160–167.

[25] Schaffer B, Davies FS, Crane JH. Responses of subtropical and tropical fruit trees to flooding in calcareous soil. Hort Sci. 2006;41:549–555.

[26] Colmer TD, Voesenek LACJ. Flooding tolerance: suites of plant traits in variable environments. Funct Plant Biol. 2009;36:665–681.

[27] García-Sanchez F, Syvertsen JP, Gimeno V, Botia P, Perez-Perez JG. Responses to flooding and drought stress by two citrus rootstock seedlings with different water-use efficiency. Physiol Plant. 2007;130:532–542.

[28] Rodríguez-Gamir J, Ancillo G, González-Mas MC, Primo-Millo E, Iglesias DJ, Forner-Giner MA. Root signalling and modulation of stomatal closure in flooded citrus seedlings. Plant Physiol Biochem. 2011;49:636–645.

[29] Arbona V, Gomez-Cadenas A. Hormonal modulation of citrus responses to flooding. J Plant Growth Regul. 2008;27:241–250.

[30] Martínez-Alcántara B, Jover S, Quiñones A, Forner-Giner MA, Rodríguez-Gamir J, Legaz F, Primo-Millo E, Iglesias DJ. Flooding affects uptake and distribution of carbon and nitrogen in citrus seedlings. J Plant Physiol. 2012;169:1150–1157.

[31] Arbona V, Hossain Z, López-Climent MF, Pérez-Clemente RM, Gómez-Cadenas A. Antioxidant enzymatic activity is linked to waterlogging stress tolerance in citrus. Physiol Plant. 2008;132:452–466.

[32] Syvertsen JP, Zablotowicz RM, Smith ML. Soil-temperature and flooding effects on 2 species of citrus. 1. Plant-growth and hydraulic conductivity. Plant Soil. 1983;72:3–12.

[33] Ponnamperuma FN. Effects of flooding on soils. In: Kozlowski TT, editor. Flooding and Plant Growth. New York: Academic Press; 1984. p. 9–45.

[34] Larson KD, Graetz D, Schaffer B. Flood-induced chemical transformations in calcareous agricultural soils of South Florida. Soil Sci. 1991;152:33–40.

[35] Gimeno V, Syvertsen JP, Simon I, Martinez V, Camara-Zapata JM, Nieves M, Garcia-Sanchez F. Interstock of 'Valencia' orange affects the flooding tolerance in 'Verna' lemon trees. HortSci. 2012;47:403–409.

[36] Martínez-Cuenca MR, Quiñones A, Primo-Millo E, Forner-Giner MA. Flooding impairs Fe uptake and distribution in citrus due to the strong down-regulation of genes involved in strategy I responses to Fe deficiency in roots. PLoS ONE. 2015;10(4):e0123644. doi:10.1371/ journal. pone.0123644.

[37] Briat J. Metal-ion-mediated oxidative stress and its control. In: Montagu M, Inze D, editors. Oxidative Stress in Plants. London: Taylor and Francis; 2002. p. 171–190.

[38] Rodríguez-Gamir J, Primo-Millo E, Forner JB, Forner-Giner MA. Citrus rootstock responses to water stress. Sci Hortic. 2010;126:95–102.

[39] Syvertsen JP, Graham JH. Hydraulic conductivity of roots, mineral nutrition, and leaf gas exchange of Citrus rootstocks. J Am Soc Hort Sci. 1985;110:865–869.

[40] Pedroso FKJV, Prudente DA, Bueno ACR, Machado EC, Riberiro RV. Drought tolerance in citrus trees is enhanced by rootstock-dependent changes in root growth and carbohydrate availability. Environ Exp Bot. 2014;101:26–35.

[41] Syvertsen JP, Lloyd JJ. Citrus. In: Schaffer BA, Andersen PC, editors. Handbook of Environmental Physiology of Fruit Crops. Vol. II. Subtropical and Tropical Crops. Boca Raton, FL: CRC Press; 1994. p. 65–99.

[42] Medina CL, Machado FC. Gas exchange and water relations of Valencia orange tree grafted on Rangpur lime and *Poncirus trifoliata* submitted to a water deficit. Bragantia, Campinas. 1998;57:15–22.

[43] Enstone DE, Peterson CA. Effects of exposure to humid air on epidermal viability and suberin deposition in maize (*Zea mays* L.) roots. Plant Cell Environ. 1998;21:837–844.

[44] North GB, Martre P, Nobel PS. Aquaporins account for variations in hydraulic conductance for metabolically active root regions of *Agave desserti* in wet, dry, and rewetted soil. Plant Cell Environ. 2004;27:219–228.

[45] Lei YB, Yin CY, Li CY. Differences in some morphological, physiological, and biochemical, responses to drought stress in two contrasting populations of *Populus przewalskii*. Physiol Plant. 2006;127:182–191.

[46] Arbona V, Iglesias DJ, Jacas J, Primo-Millo E, Talon M, Gomez-Cadenas A. Hydrogel substrate amendment alleviates drought effects on young citrus plants. Plant Soil. 2005;270:73–82.

[47] Gonzalez-Mas MC, Llosa MJ, Quijano A, Forner-Giner MA. Rootstock effects on leaf photosynthesis in 'Navelina' trees grown in calcareous soil. HortSci. 2009;44:208–283.

[48] Jover S, Martinez-Alcántara B, Rodríguez-Gamir J, Legaz F, Primo-Millo E, Forner J, Forner-Giner MA. Influence of rootstocks on photosynthesis in navel orange leaves: effects on growth, yield and carbohydrate distribution. Crop Sci. 2012;52:836–848.

[49] Mehouachi J, Serna D, Zaragoza S, Agusti M, Talon M, Primo-Millo E. Defoliation increases fruit abscission and reduces carbohydrate levels in developing fruits and woody tissues of *Citrus unshiu*. Plant Sci. 1995;107:189–197.

[50] Iglesias DJ, Lliso I, Tadeo FR, Talon M. Regulation of photosynthesis through source: sink imbalance in citrus is mediated by carbohydrate content in leaves. Physiol Plant. 2002;16:563–572.

[51] Rodriguez-Gamir J, Intrigliolo DS, Primo-Millo E, Forner-Giner MA. Relationships between xylem anatomy, root hydraulic conductivity, leaf/root ratio and transpiration in citrus trees on different rootstocks. Physiol Plant. 2010b;139:159–169.

[52] Cameron J, Soost RK. C35 and C32: citrange rootstocks for citrus. HortSci. 1986;21:157–158.

[53] Bitters WP. Reaction of some new citrus hybrids and citrus introductions as rootstocks to inoculations with tristeza virus in California. In: Proceedings 5th International Organization of Citrus Virology; 1972, pp. 112–120.

[54] Geraci G, Lo Giudice V, Inserra RN. Response of Citrus spp. and hybrid rootstocks to *Tylenchulus semipenetrans*. Riv Ortoflorofrutt. 1981; 65:169–172.

[55] Young R, Hearn CJ. Screening citrus hybrids for cold hardiness. HortSci. 1972;1:14–18.

[56] Forner JB, Forner-Giner MA, Alcaide A. Forner-Alcaide 5 and Forner-Alcaide 13: two new citrus rootstocks released in Spain. HortSci. 2003;38:629–630.

[57] Forner-Giner MA, Alcaide A, Primo-Millo E, Forner JB. Performance of 'Navelina' orange on 14 rootstocks in Northern Valencia (Spain). Sci Hortic. 2003;98:223–232.

[58] Webster AD. Rootstock and interstock effects on deciduous fruit tree vigour, precocity, and yield productivity. NZJ Crop Hortic Sci. 1995;23:373–382.

[59] Castle WS, Baldwin JC, Muraro RP. Rootstocks and the performance and economic returns of 'Hamlin' sweet orange trees. HortSci. 2010;45:875–881.

[60] Cheng FS, Roose ML. Origin and inheritance of dwarfing by the Citrus rootstock *Poncirus trifoliata* 'Flying Dragon'. J Am Soc Hortic Sci. 1995;120:286–291.

[61] Legua P, Forner JB, Hernández F, Forner-Giner MA. Physicochemical properties of orange juice from ten rootstocks using multivariate analysis. Sci Hortic. 2013;160:268–273.

[62] Legua P, Forner JB, Hernández F, Forner-Giner MA. Total phenolics, organic acids, sugars and antioxidant activity of mandarin (*Citrus clementina* Hort. ex Tan.): variation from rootstock. Sci Hortic. 2014;174:60–64.

[63] Forner-Giner MA, Rodriguez-Gamir J, Martinez-Alcantara B, Quiñones A, Iglesias DJ, Primo-Millo E, Forner J. Performance of Navel orange trees grafted onto two new dwarfing rootstocks (Forner-Alcaide 517 and Forner-Alcaide 418). Sci Hortic. 2014;179:376–387.

[64] Martínez-Alcántara B, Rodriguez-Gamir J, Martínez-Cuenca MR, Iglesias DJ, Primo-Millo E, Forner-Giner MA. Relationship between hydraulic conductance and citrus dwarfing by the Flying Dragon rootstock (*Poncirus trifoliata* L. Raft var. monstruosa). Trees. 2013;27:629–638.

[65] Basile B, Marsal J, DeJong TM. Daily shoot extension growth of peach trees growing on rootstocks that reduce scion growth is related to daily dynamics of stem water potential. Tree Physiol. 2003;23:695–704.

[66] Georgiou A. Evaluation of rootstocks for 'Clementine' mandarin in Cyprus. Sci Hortic. 2002;93:29–38.

[67] Georgiou A. Evaluation of rootstocks for the Cyprus local lemon variety 'Lapithkiotiki'. Sci Hortic. 2009;123:184–187.

[68] Bassal MA. Growth, yield and fruit quality of 'Marisol' clementine grown on tour rootstocks in Egypt. Sci Hortic. 2009;119:132–137.

[69] Tworkoski T, Fazio G. Hormone and growth interactions of scions and size-controlling rootstocks of young apple trees. Plant Growth Regul. 2016;78:105–119.

[70] Da Silva SR, Stuchi ES, Girardi EA, Cantuarias-Aviles T, Bassan MM. Performance of 'Span Americana' mandarin on different rootstocks. Rev Bras Frutic. 2013;35:1052–1058.

[71] Westwood MN, Roberts AN. The relationship between trunk cross-sectional area and weight of apples trees. J Am Soc Hortic Sci. 1970;95:28–30.

[72] Yildirim B, Yesiloglu T, Kamiloglu MU, Incesu M, Tuzcu O, Cimen B. Fruit yield and quality of Santa Teresa lemon on seven rootstocks in Adana (Turkey). Afr J Agric Res. 2010;5:1077–1081.

[73] Legua P, Bellver R, Forner J, Forner-Giner MA. Plant growth, yield and fruit quality of 'Lane Late' navel orange on four citrus rootstocks. Span J Agric Res. 2011;9:271–279.

[74] Roose ML, Cole PA, Atkin D, Kupper RS. Yield and tree size of four citrus cultivars on 21 rootstocks in California. J Am Soc Hort Sci. 1989;114:678–684.

[75] Perez-Perez JG, Porras I, Garcia-Lidon A, Botia P, Garcia-Sanchez F. Fino lemon clones compared with the lemon varieties Eureka and Lisbon on two rootstocks in Murcia (Spain). Sci Hortic. 2005;106:530–538.

[76] Forner-Giner MA, Hueso JJ, Agüera JM, Legua P, Forner JB. Effects of citrus rootstocks on growth, yield and fruit quality of Navelate orange. J Food Agric Environ. 2011;9:400–403.

[77] Legua P, Bellver R, Forner JB, Forner-Giner MA. Trifoliata hybrids rootstocks for 'Lane Late' navel orange in Spain. Sci Agric. 2011;68:548–553.

[78] Forner-Giner MA, Hueso JJ, Agüera JM, Forner JB. Performance of 'Clausellina' mandarin on four rootstocks. J Food Agric Environ. 2010;8:328–331.

[79] Al-Jaleel A, Zekri M, Hammam Y. Yield, fruit quality, and tree health of 'Allen Eureka' lemon on seven rootstocks in Saudi Arabia. Sci Hortic. 2005;105:457–465.

[80] Figueiredo JO, Negri JD, Mattos Jr. D, Pio RM, Laranjeira FF, Garcia VXP. Behaviour of fourteen rootstocks for eureka lemon [*Citrus limon* (L) Burm. F.] cv. Km 47, in Araraquara, SP, Brazil. Rev Bras Frutic. 2005;1:73–76.

[81] Tuzcu Ö, Yildrim B, Yesiloglu T. Effect of different rootstocks and sectors on fruit yield and its distribution depending to the tree canopy. In: Proceedings of the International Society of Citriculture, Xth Congress, Agadir, Morocco; 2004, pp. 405–408.

[82] Mourão Filho FAA, Espinoza-Núñez E, Stuchi ES, Ortega EMM. Plant growth, yield and fruit quality of 'Fallglo' and 'Sunburst' mandarins on four rootstocks. Sci Hortic. 2007;114:45–49.

[83] Georgiou A, Gregoriou C. Growth, yield and fruit quality of 'Shamouti' orange on fourteen rootstocks in Cyprus. Sci Hortic. 1999;80:113–121.

[84] Blanco MS, Torrecillas A, León A, Del Amor F. The effect of different irrigation treatments on yield and quality of Verna lemon. Plant Soil. 1989;120:299–302.

[85] Gregoriou C, Economides CV. Tree growth, yield and fruit quality of 'Ortanique' tangor on eleven rootstocks in Cyprus. J Am Soc Hort Sci. 1993;118:335–338.

[86] Olmo M, Nadas A, García JM. Non destructive methods to evaluate maturity level of oranges. J Food Sci. 2000;65:365–369.

[87] Lee HS, Castle WS. Seasonal changes of carotenoid pigments and colour in Hamlin, Earlygold, and Budd Blood orange juices. J Agric Food Chem. 2001;49:877–882.

[88] Economides CV, Gregoriu C. Growth, yield and fruit quality of nucellar frost 'Marsh' grapefruit on fifteen rootstocks in Cyprus. J Am Soc Hort Sci. 1993;118:326–329.

[89] Misra KK, Sigh R, Jaiswal HR. Effect of rootstocks on the fruit quality of lemon (*Citrus limon*). Indian J Agric Sci. 1999;69:670–671.

[90] Zekri M, Al-Jaleel A. Evaluation of rootstocks for Valencia and Navel orange trees in Saudi Arabia. Fruits. 2004;59:91–100.

[91] Davies FS, Albrigo LG. Citrus. Wallingford, UK: CAB International; 1994.

[92] Verdú EL. Clemenules behavior on tolerant rootstocks. In: I Plana Citriculture Congress 1993, pp. 26–27.

[93] Demirkeser TH, Kaplankiran M, Yildiz E. The effects of some citrus rootstocks on fruit yield and quality for Rohde Red Valencia orange during the period of the juvenility in Dortyol (Hatay-Turkey) conditions. In: 7th International Society of Citrus Nurserymen Congress, Cairo, Egypt, Abst. No. 28; September 2005.

[94] Nisperos-Carriedo MO, Baldwin EA, Moshonas MG, Shaw PE. Determination of volatile flavor components, sugars, and ascorbic, dehydroascorbic, and other organic acids in calamondin (*Citrus mitis* Blanco). J Agric Food Chem. 1992;40:2464–2466.

[95] Kelebek H, Selli S, Canbas A, Cabaroglu T. HPLC determination of organic acids, sugars, phenolic compositions and antioxidant capacity of orange juice and orange wine made from a Turkish cv. Kozan. Microchem J. 2009;91:187–192.

[96] Kaplankiran M, Demirkeser TH, Yildiz E. The effects of some citrus rootstocks of fruit yield and quality for Okitsu Satsuma during the juvenility period in Dortyol (Hatay, Turkey) conditions. In: 7th International Society of Citrus Nurserymen Congress, Cairo, Egypt; 2005, p. 27.

[97] Albertini MV, Carcouet E, Pailly O, Gambotti C, Luro F, Berti L. Changes in organic acids and sugars during early stages of development of acidic and acidless citrus fruit. J Agric Food Chem. 2006;54:8335–8339.

[98] Wutscher HK, Maxwell NP, Shull AV. Performance of nucellar grapefruit, *Citrus paradise* Macf., on 13 rootstocks in South Texas. J Am Soc Hort Sci. 1975;100:48–51.

[99] Continella G, Germana C, La Rosa G, Tribulato E. Performance and physiological parameters of 'Comune' Clementine influenced by four rootstocks. In: Proceedings VI International Citrus Congress, Tel Aviv (Israel); March 1998, pp. 91–100.

[100] Fallahi E, Rodney DR. Tree size, yield, fruit quality, and leaf mineral nutrient concentration of 'Fairchild' mandarin on six rootstocks. J Am Soc Hort Sci. 1992;117:28–31.

Chapter 9

Root Growth, Morphological and Physiological Characteristics of Subtropical and Temperate Vegetable Crops Grown in the Tropics Under Different Root-Zone Temperature

Jie He

Additional information is available at the end of the chapter

http://dx.doi.org/10.5772/64824

Abstract

Root growth and morphology are important for maximizing water uptake and mineral absorption. Similar to the plants grown in the soil, in a soilless culture such as an aeroponic system, the amount of water and nutrient available to a plant is determined by the root surface area and volume of nutrient solution with which its roots are in contact. Furthermore, plant roots can alter their nutrient acquisition capacity by adjusting their morphological and/or physiological characteristics to meet changes in shoot nutrient demand in response to environmental stress. Subtropical and temperate vegetables have successfully been grown aeroponically in the tropics by simply cooling the root zone (RZ) while their aerial portions are subjected to fluctuating atmospheric temperatures. This paper focused on RZ temperature (RZT) on root and shoot growth, and root morphology of subtropical and temperate vegetable crops grown in the tropics. The impacts of RZT on water relations as well as nitrate (NO_3−) uptake and assimilation of these vegetable crops were also discussed.

Keywords: nitrate uptake and assimilation, root morphology, root-zone temperature, water relations

1. Introduction

Plants are photoautotrophic and manufacture their own food through the process of photosynthesis. However, they must acquire water and minerals from the environment for photosynthesis to occur. While CO_2 comes from the atmosphere for photosynthesis, plants obtained water and minerals from the soil or other growth media via their roots. Plants are sessile and adapt their morphological structures to the encountered environmental conditions. The morphology of the root system such as root length, number of root tip, root diameter, root surface area, and root volume varies greatly depending on the plant species, soil composition, and water and mineral nutrients availability [1, 2] and root-zone temperature (RZT) [2, 3]. Root plasticity provides the sessile plants to adjust their structure to environmental conditions as they change [1]. Changes in root morphology are important for maximizing water uptake [1, 3, 4] and mineral absorption to alleviate stress of nutrient deficiency [4–7]. Furthermore, plant roots alter their nutrient acquisition capacity not only by adjusting the morphological characteristics but also the physiological performance such as nitrate (NO_3^-) uptake and assimilation to meet changes in shoot nutrient demand in response to environmental stress [2, 4].

Root morphology and physiology are closely associated with the growth of the aerial parts and shoot productivity [2, 3, 8]. For instance, plant growth and productivity are often limited by high temperature and thus restrict the growth of subtropical and temperate crops in the tropics. However, since 1997, our research team has successfully grown certain subtropical and temperate crops in Singapore by simply cooling their roots while their aerial portions are subjected to hot fluctuating atmospheric conditions [9, 10]. Working on subtropical and temperate vegetable crops grown in the tropics, in this paper, the author first focused on effects of RZT on root morphology. The impacts of RZT on water uptake and water relations as well as NO_3^- uptake and assimilation were also discussed.

2. Effects of root-zone temperature on root morphology

Root systems have higher ratios of surface area to volume that effectively explore a larger volume of soil [11]. Similar to the plants grown in the soil, in aeroponic systems, root surface area determines the amount of water and mineral uptake [2, 12, 13]. As root systems are responsible for acquisition of water and mineral nutrients, it is not surprising that root morphology is highly influenced by rhizosphere environments such as RZT [2].

When the plant is exposed to high temperatures, root development is adversely affected [14]. However, manipulating RZT alone has a great influence on root development, perhaps even greater than shoot temperature manipulation [15–18]. Because of the changes of RZT, there is great variation of root morphology, especially root length which is a more sensitive indicator of RZT impact compared to root biomass. For example, grown underoptimum RZT of 30°C, root length of rape (*Brassica napus* cv. "Emeralk") was fivefold longer compared to those grown under superoptimum RZT of 35°C. However, there was only twofold difference in root biomass, which was found between the plants grown under two different RZTs. Higher root length/weight biomass ratio was due to the smaller diameter of the successive orders of lateral roots [15]. It was reported that in both heat tolerant and sensitive clones of potato (*Solanum tuberosum* L), smaller number and shorter length of lateral roots were observed at 30°C RZT than 20°C RZT resulting from a reduction of cell division rate and followed by cessation of cell elongation in roots [16]. It was found that maximum seminal root elongation and first-order lateral root initiation and elongation of Sorghum (*Sorghum bicolo*), occurred at 25°C but they were severely inhibited at 40°C [17].

We have found that subtropical and temperate vegetables grown in the tropical greenhouse under cool RZTs (C-RZTs) established much bigger root systems compared to those subjected to hot ambient RZT (A-RZT) [2]. For instance, the total root length of capsicum plants (*Capsicum annuum* Indra F1-hybrid) at C-RZT increased about 400 cm in 2 weeks during the experimental period (**Figure 1A**). When plants were transferred from C-RZT to A-RZT (C → A-RZT), increases in their total root length seemed to stop compared to C-RZT plants [19]. Capsicum plants grown at A-RZT did not show any increases and had the lowest total root length during the entire experimental period. However, after transferring capsicum plants from A-RZT to C-RZT (A → C-RZT), linear increase in total root length was observed.

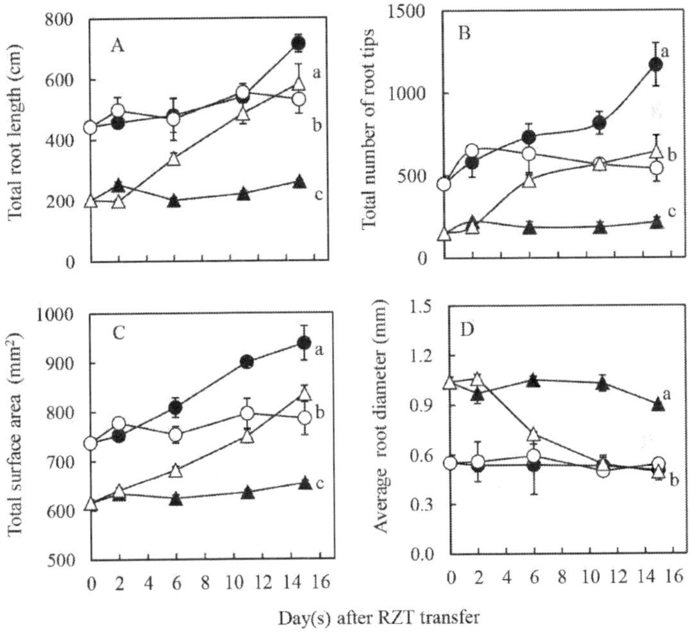

Figure 1. Total root length (A), total number of root tips (B), total surface area (C), and average root diameter (D) of capsicum (*Capsicum annuum* Indra F1-hybrid) during different RZT treatments: C-RZT (•), A-RZT (▴), C → A-RZT (○), and A → C-RZT (Δ). RZT treatments were started 3 weeks after transplanting. Means with different letters are statistically different ($P < 0.05$; $n = 6$) as determined by Tukey's multiple comparison test. Redrawn from Tan [19].

Differences in total number of root tip (**Figure 1B**) and total surface area (**Figure 1C**) among the four different RZT treatments were similar to those of total root length (**Figure 1A**). Compared to plants grown at A-RZT, the average root diameter of C-RZT plants was much smaller (**Figure 1D**). It is surprising to note that there were no increases in root diameter after transferring plants from C-RZT to A-RZT (C → A-RZT) as capsicum plants were similar to those of C-RZT plants throughout the experiment. However, A → C-RZT transfer resulted in the development of new and finer roots in A → C-RZT plants. Thus, the average root diameter of A → C-RZT capsicum was significantly smaller than A-RZT by day 6 after RZT transfer (**Figure 1D**). Similar to capsicum plants, lettuce grown at C-RZT showed linear increase in total root length over the 2-week period (**Figure 2A**). Unlike the capsicum plants (**Figure 1A**), increase in total root length was also observed in C → A-RZT lettuce plant and the increase rate was similar to C-RZT lettuce plants from day 0 to day 8 after RZT transfer. However, C → A-RZT lettuce plants maintained their total root length of about 600 cm from day 10 to day 14 after RZT transfer. There was no significant increase in total root length of A-RZT lettuce over the 2-week experimental period. Again, unlike the capsicum plants (**Figure 1A**), A → C-RZT lettuce exhibited increase in total root length only after 8 days of RZT transfer (**Figure 2A**). The total root tip number of C-RZT lettuce was consistently higher than lettuce grown at other RZTs (**Figure 2B**). Unlike the capsicum plants (**Figure 1B**), there was an increase in total root tip number for C → A-RZT lettuce although the net increase decreased significantly

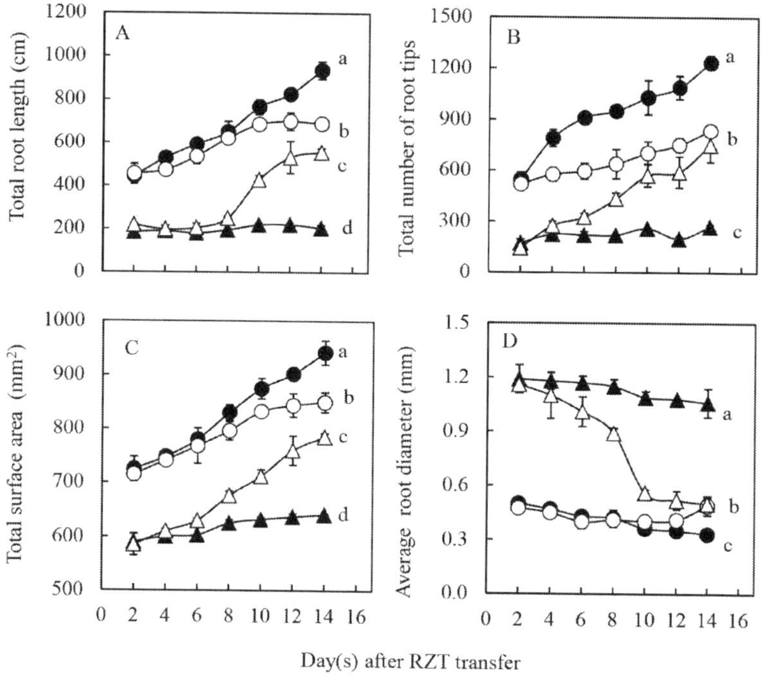

Figure 2. Total root length (A), total number of root tips (B), total surface area (C), and average root diameter (D) of lettuce (*Lactuca sativa* L. "Panama") during different RZT treatments: C-RZT (•), A-RZT (▴), C → A-RZT (○), and A → C-RZT (Δ). RZT treatments were started 3 weeks after transplanting. Means with different letters are statistically different ($P < 0.05$; $n = 6$) as determined by Tukey's multiple comparison test. Redrawn from Tan [19].

after C → A-RZT transfer (**Figure 2B**). The total root tip number of A-RZT lettuce did not change significantly throughout the whole experiment. When lettuce was transferred from A → C-RZT, there was rapid increase in the total root tip number. The root tip number of A → C-RZT lettuce became significantly higher than A-RZT lettuce by day 6 after RZT transfer. At day 14 after RZT transfer, total root tip number of A → C-RZT lettuce was 2.8 times higher than A-RZT plants. Responses of total surface area of lettuce plants to different RZTs (**Figure 2C**) were similar to those of total root length (**Figure 2B**). Similar to those of capsicum plants (**Figure 1D**), lettuce grown at A-RZT had the highest average root diameter while C-RZT plants had the least (**Figure 2D**). The average root diameter of C → A-RZT lettuce remained similar to C-RZT plants during the first 10 days of RZT transition. However, unlike capsicum plants (**Figure 1D**), there was slight root thickening in C → A-RZT lettuce roots after 2 weeks of RZT transfer. By day 14 after RZT transfer, the average root diameter of C → A-RZT lettuce was much thicker than C-RZT plants. Similar to those of capsicum plants, A → C-RZT lettuce, developed new and finer root (**Figure 2D**).

Root morphological analysis of both subtropical and temperate vegetables revealed that high RZT inhibited root elongation, branching, and hair formation but increased root diameter. These were also observed in other plant species [16, 17]. However, effects of RZT transfer on morphology were different between capsicum and lettuce plants (**Figures 1** and **2**). These finding suggest that effects of RZT on root morphology is species-dependent. It was also reported that root length and diameter appeared to be inversely related to a study using *Secale cereale* seedlings [20]. Root thickening, or an increase in diameter, was controlled through signals emanating from shoot apices and root tips [21]. Root thickening may also be accompanied by associated changes in microfibril angles within expanding cell walls [21]. The chemical signals involved in root thickening may be ethylene [22]. The role of ethylene in inhibition of root elongation and root thickening was further confirmed by our team [13]. Our recent study with a recombinant inbred line (RIL) of lettuce and its parental lines (*Lactusa serriola* × *Lactuca sativa* "Salinas") that were grown in a tropical greenhouse under 24°C-RZT and hot A-RZT showed that higher RZ ethylene concentrations accumulated in A-RZT plants compared to that of 24°C-RZT plants. Lowest RZ ethylene concentration corresponded with highest shoot fresh weight [23]. Our results indicated that the presence of an ethylene inhibitor promoted root elongation at high RZT of 38°C. Without ethylene inhibitor, root elongation at high RZT was significantly inhibited. Our previous ^{14}C feeding experiments implied that both capsicum and lettuce grown under C-RZT indeed had higher assimilation rates of ^{14}C and their younger developing leaves exhibited greater sink strength [24]. Further studies concluded that plant growth was the result of interaction between source leaves and carbon partitioning among competitive sinks [25–27]. It was interesting to note that A-RZT lettuce had higher fresh weight and dry weight root/shoot ratio than that of C-RZT lettuce [19, 23]. This suggested that more photoassimilate may be distributed to the lettuce roots under hot A-RZT conditions. The RZT transfer experiments confirmed that A-RZT induced greater levels of ^{14}C delivered to the lettuce root system. However, it was surprising that the high ^{14}C translocated to the roots of A-RZT lettuce which was not accompanied by a greater root development [19, 24]. This may be attributed to the higher respiration rates in the roots which may require energy for the active uptake of water and nutrients in a poorly developed root system [26, 28]. The high rate of respiration may have taken place at the expense of root development. It is highly likely that there exists a negative feedback mechanism among root respiration, water uptake, nutrient absorption, root morphology, and high RZT. Our results showed that capsicum grown at C-RZT had lower root/shoot ratios than A-RZT plants [19]. C-RZT resulted in longer total root length, greater total root tip number, larger root surface area, and smaller root diameter in both lettuce and capsicum plants. However, the trend of the capsicum root/shoot ratios was different from that of lettuce because the capsicum grown at C-RZT had higher root/shoot ratio than A-RZT capsicum [19]. C → A-RZT transfer also caused a decline in root/shoot ratio while A → C-RZT resulted in an increase in root/shoot ratios of capsicum. These suggested more photo-assimilate partitioning to roots of C-RZT capsicum and less so for A-RZT capsicum. This was further supported by the ^{14}C feeding experiments which showed that more ^{14}C was found in the roots of C-RZT capsicum compared to the roots of A-RZT capsicum. The transfer from C → A-RZT also resulted in decreased ^{14}C found in C → A-RZT capsicum roots. The reverse was demonstrated in A → C-RZT capsicum which had higher root ^{14}C than A-RZT plants. Therefore, lettuce grown at A-RZT distributed more ^{14}C to their roots while capsicum grown at C-RZT distributed more ^{14}C to the roots [24].

3. Effects of root-zone temperature on water relations

Changes of root morphology are important for maximizing water and dissolve mineral uptake [1, 11, 29–31]. Plant roots can alter not only their morphological but also their physiological characteristics to meet changes in shoot water and nutrient demand in response to environmental stress [29]. Manipulation of RZT resulted in changes of root morphology, water, and mineral uptake and transport [12, 32–35]. For plants grown in soil, high RZT causes not only poor root growth and development [36] but also results in spatial water and mineral nutrient availability in soil [37] and reduces uptake efficiency per unit root length [36, 37]. In our aeroponic system, plant roots are continuously sprayed with nutrient mist, and there is no spatial variation in water and mineral nutrient availability. However, water deficits and mineral deficiency occurred at high A-RZT due to poor root system development and lower rate of water uptake [9, 10, 12].

We have previously reported that supra-optimal A-RZT caused shoot water deficit by altering the balance between water uptake by the root system and water loss from the shoot [38, 39]. Water deficits resulting in stomata closure measured by stomatal conductance are due to the direct effect of reduced shoot water potential (ψ_{shoot}) and leaf relative water content (RWC) [38–42]. **Table 1** summaries ψ_{shoot} of subtropical vegetable crops of capsicum (*C. annuum* Indra F1-hybrid) and Chinese broccoli (*Brassica alboglabra* Bailey) [38] grown at 25°C-RZT and A-RZT while their shoots were maintained at fluctuating ambient temperatures. **Table 1** shows predawn and midday ψ_{shoot} were higher in 25°C-RZT than A-RZT plants in both vegetable species. Leaf RWC were further determined in both subtropical (capsicum and Chinese broccoli) and temperate vegetables (lettuce). Similar to the results of ψ_{shoot}, RWC was significantly lower in all plants grown at A-RZT that at C-RZT measured predawn and midday (**Table 2**).

Vegetable species	25°C-RZT	A-RZT
Capsicum		
Predawn ψ_{Leaf} (Mpa) (n = 6) *	−0.51 ± 0.070	−2.95 ± 0.18
Midday ψ_{Leaf} (Mpa) (n = 6) *	−1.45 ± 0.092	−4.87 ± 0.24
Chinese broccoli		
Predawn ψ_{Leaf} (Mpa) (n = 6) *	−0.21 ± 0.03	−1.07 ± 0.06
Midday ψ_{Leaf} (Mpa) (n = 6) *	−0.85 ± 0.07	−2.03 ± 0.12

* Significant interaction between the two RZTs at $P < 0.01$.

Table 1. ψ_{Leaf} of subtropical vegetable crops of capsicum [19] and Chinese broccoli (*Brassica alboglabra* Bailey) [38] grown at two different RZTs while their shoots were maintained at fluctuating ambient temperatures under 100% prevailing solar radiation. All the measurements were done after 30 days of transplanting. Values shown are means ± standard deviation.

Vegetable species	Cool-RZT*	A-RZT
Capsicum		
Predawn RWC (%) (n = 6)[#]	93.2 ± 0.22	80.3 ± 0.28
Midday RWC (%) (n = 6)[#]	81.3 ± 0.31	51.7 ± 0.29
Chinese broccoli		
Predawn RWC (%) (n = 6)[#]	97.6 ± 0.23	85.4± 0.16
Midday RWC (%) (n = 6)[#]	85.5 ± 0.27	64.9 ± 0.12
Lettuce		
Predawn RWC (%) (n = 6)[#]	95.1 ± 0.17	88.8 ± 0.35
Midday RWC (%) (n = 6)[#]	89.2 ± 0.29	60.1 ± 0.46

* Cool-RZT, 25°C-RZT for capsicum and Chinese broccoli; 20°C-RZT for lettuce.
[#] Significant interaction between the two RZTs at $P < 0.01$.

Table 2. Leaf RWC of subtropical capsicum (*Capsicum annuum* Indra F1-hybrid) [19] Chinese broccoli (*Brassica alboglabra* Bailey) [19], and temperate vegetable crops of lettuce [39] grown at two different RZTs while their shoots were maintained at fluctuating ambient temperatures under 100% prevailing solar radiation. All the measurements were done after 30 days of transplanting. Values shown are means ± standard deviation.

Based on the results shown in **Tables 1** and **2**, it is obviously that subtropical and temperate vegetables grown in the tropical greenhouse had experienced permanent water deficit (reduced predawn ψ_{shoot} or/and leaf RWC) and midday (lower midday ψ_{shoot} or/and leaf RWC) when they were grown at A-RZT. In another experiment with capsicum, leaf g_s, root hydraulic conductivity, and shoot ψ_{shoot} declined after transferring plants from 20°C-RZT to A-RZT [42]. It was explained that supraoptimal RZTs caused a reduction in root hydraulic conductivity and might lower ψ_{shoot}, which in turn could cause stomatal closure [42]. However, water deficit in 20 or 25°C-RZT plants was alleviated due to the larger root system [19].

4. Effects of root-zone temperature on NO_3– uptake and assimilation

NO_3– is the major N source available in aerobic soils [43]. Once uptaken by root cells, NO_3– can be redirected out of the root cell, either by extrusion into the external medium or by unloading into the xylem vessels to reach the aerial organs [29]. The third possible fate for NO_3–, in roots as well as in leaves, is its uptake by the vacuole where it participates in the general osmoticum or serves as a reservoir to sustain the growth process when the external nitrogen supply becomes limiting [44].

The effects of RZT on NO_3– and N contents in plants have been reported [45–47]. N contents were reduced in plants grown under high RZT [39, 48–50]. For instance, Du and Tachibana [48] grew cucumber (cv. "Sharp I") plants hydroponically at several RZTs: 25 (control), 30, 35, or 38°C, with shoot's temperature at 26/23 ± 3°C (day/night). Total N concentration in leaf was reduced as the RZT was raised to 35°C and to 38°C in particular. Similar results were obtained our research team with lettuce (cv. "Palma") plants [39]. Leaf organic N content was 32% lower in A-RZT plants than 20°C-RZT plants. Transfer of plants between these two RZTs altered leaf N content after 6 days. Leaf N content increased in A → 20°C-RZT plants, while decreased in 20°C → A-RZT plants. After 10 days of reciprocal RZT transfer, A → 20°C-RZT plants and 20°C → A-RZT plants had a similar leaf N content [39]. The results also showed tight temporal coupling of leaf N content, and light- and CO_2- saturated photosynthetic O_2 evolution rate throughout the reciprocal temperature transfers. He et al. [39] suggested that the decreased nutrient status of lettuce plants caused nonstomatal limitation of photosynthesis under high A-RZT conditions. In lettuce (cv. "Panama"), Tan et al., [12] reported that 20°C-RZT plants had higher leaf N concentrations on the basis of per unit dry weight compared with plants grown at A-RZT. Total shoot and root NO_3 of 20°C-RZT plants were higher than A-RZT plants. 20°C → A-RZT plants suffered from a reduction of total mineral accumulation, and A → 20°C-RZT plants increased in total mineral accumulation [12]. In another study, Yeager et al. grew *Ilex crenata* Thunb. "Rotundifolia" plants in sand culture with the RZTs at 28, 34, or 40°C for 6 h daily. They found that root and shoot N accumulation (mg N/g dry weight) decreased when RZTs were increased from 28 to 40°C. These plants were fertilized twice daily with 500 mL of either 75, 150, or 225 mg N/L, to determine applied N rate on growth and N accumulation of "Rotundifolia" holly. The results showed that root and shoot N accumulation depended on RZT and the N rate. N accumulated by roots and shoots increased when the applied N concentration increased from 75 to 225 mg/L at each RZT. However, root and shoot N accumulation decreased with the root zone at 40°C compared with 28°C for 75 and 225 mg N/L applied. These data indicated that increased N fertilization rates would not alleviate growth reductions of holly caused by high RZT (40°C) [49].

In contrast, high RZT increased N content in plants has also been reported [49–51]. Johnson and Ingram exposed root systems of *Pittosporum tobira* Thunb plants to temperatures of 27, 30, or 40°C for 6 h daily for 7 months with air temperature around 30/24°C (day/night). They found that N level in leaf tissue (newly expanded leaves) was increased in plants at 40°C medium temperature [50]. This was supported by the work of Gosselin and Trudel who transferred the 10-week-old pepper (cv. "Bell Boy") plants to five different RZTs (12, 18, 24, 30, or 36 ± 2°C)

http://dx.doi.org/10.5772/64824

with air temperature maintained at 22–24/18–20°C (day/night), for a period of 8 weeks. Their results showed that leaf N concentration increased with the increase of RZT from 12 to 36°C [51]. Cruz et al. grew Carob (*Ceratonia siliqua* L.) seedlings at different RZTs (10–40°C) with shoot temperature at 24 ± 1/20 ± 1°C (day/night). When nitrate (3 mM) was supplied as nitrogen source of nutrient solutions, they found that organic N concentration in roots increased with the increase of RZT from 10 to 40°C. Total N concentration in both shoot and toot increased with the increased RZT from 10 to 25°C. Investigators explained that increasing RZTs induced increased in ion-uptake rates, mainly nitrogen, might be responsible for the higher N content in plant. This situation increased also the demand for carbon in the root of Carob plant [52]. As discussed earlier, the amount of mineral nutrient especially nitrogen available to a plant is determined by the root morphological structure [3, 12, 13]. The amount of mineral nutrient uptake and transport also depend on transpiration rate as long-distance movement of nutrient through plants is predominantly by bulk flow in the xylem [53].

For most higher plants, NO_3– assimilation is the major pathway by which inorganic N is converted to an organic form [54]. The conversion of NO_3– to NH_4+, which can be directly used to synthesize organic N containing compounds, is a reduction process that occurs in two steps. NO_3– is first reduced to NO_2– in the cytosol by nitrate reductase (NR). NO_2– is then translocated to the chloroplast where it is reduced into NH_4+ by nitrite reductase (NiR). NR is NO_3– inducible enzyme. Transcription of NR genes is induced by NO_3– [54–56]. The reduction of NO_3– could take place either in roots or in leaves or both [54, 56, 59]. Significant translocation of NO_3– to the shoot would occur only when the net NO_3– uptake rate was fast enough to saturate the reduction process in the roots.

In our study of subtropical vegetable Nai Bai (*Brachyponera chinensis* L.) plant and Baby butterhead lettuce (*L. sativa* L.) plants, NO_3– concentrations of leaf and root and maximum nitrate reductase activity (NRA) of leaf and roots were determined after 10, 20, and 30 days of transplanting (**Figures 3** and **4**).

For Nai Bai plants, the leaf NO_3– concentrations of either 25°C-RZT or A-RZT plants were similar at 10, 20, and 30 days after transplanting (DAT). However, the leaf NO_3– concentrations of A-RZT plants were significantly lower than those of 25°C-RZT plants (**Figure 3A**). Regardless of growth stage, 25°C-RZT plants had significantly higher root NO_3– concentrations than those of A-RZT plants (**Figure 3C**). Root NO_3– concentrations of A-RZT remained constant at all growth stages (**Figure 3C**). However, 25°C-RZT plants had lower root NO_3– concentrations at 10 DAT compared to those at 20 and 30 DAT (**Figure 3C**). At 10 DAT, leaf NRA of 25°C-RZT plants was significantly lower than those of A-RZT plants (**Figure 3B**). However, it increased from 10 to 20 DAT, and remained constant from 20 to 30 DAT. Leaf maximum NRA of A-RZT plants decreased from 10 to 20 DAT, and remained constant from 20 to 30 DAT. Therefore, at 20 and 30 DAT, maximum leaf NRA of 25°C-RZT plants was much higher than those of A-RZT plants (**Figure 3B**). Root maximum NRA of 25°C-RZT and A-RZT plants decreased from 10 to 20 DAT, and then remained constant to the 30 DAT. At 10 DAT, no significant difference in maximum root NRA was measured between 25°C-RZT and A-RZT plants. However, at 20 and 30 DAT, maximum root NRA of 25°C-RZT plants was significantly higher than those of A-RZT plants (**Figure 3D**).

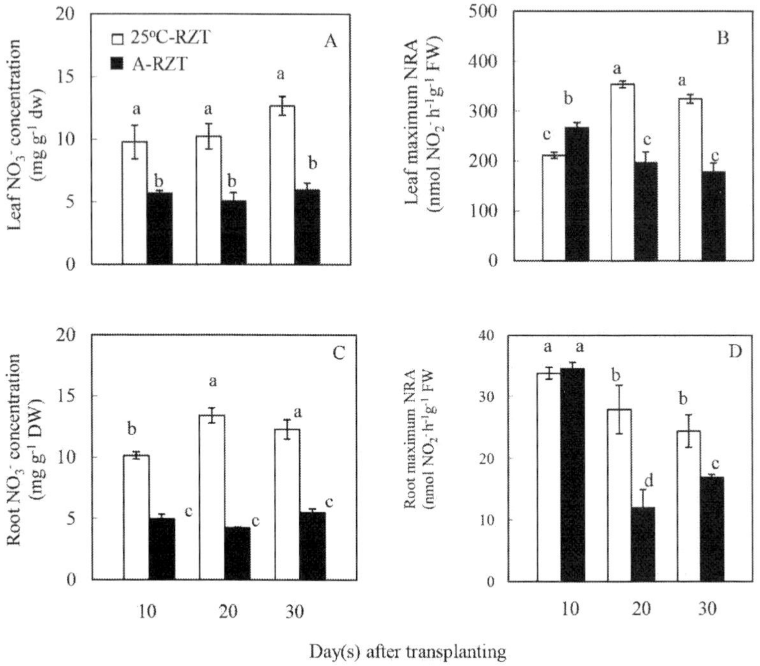

Figure 3. Leaf NO_3^- concentration (A) and maximum NRA (B), root NO_3^- concentration (C) and maximum NRA (D) of Nai Bai (*Brachyponera chinensis* L.) plant. Each point is the mean of five measurements of five different leaves from two different bins. Vertical bars represent the standard errors. Means with different letters above the columns are statistically different ($P < 0.001$) as determined by Tukey's multiple comparison test (unpublished data).

The leaf NO_3– concentrations of 20°C-RZT were significantly higher compared to those of A-RZT plants, at different growth stages (**Figure 4A**). With prolonged growth to 30 DAT, leaf and root NO_3– concentrations of 20°C-RZT plants increased (**Figure 3A**). However, there was no significant difference in leaf and root NO_3– concentrations at A-RZT among the three growth stages (**Figure 4C**). Leaf maximum NRA in 20°C-ZT and A-RZT plants was highest at 10 DAT and decreased at 20 DAT and further decreased at 30 DAT (**Figure 4A**). Regardless of growth stage, leaf maximum NRA was significantly higher in 20°C-ZT than in A-RZT plants (**Figure 4B**). It was surprise to observe that root maximum NRA was many times higher in A-RZT plants than in 20°C-RZT plants, indicating that RZT altered the site of NO_3– assimilation (**Figure 4D**).

The reduction of NO_3– could take place either in roots, in leaves, or in both [57, 58]. For Baby butterhead plants, hot A-RZT treatment switched NO_3– reduction from shoot to root, evidenced by the higher NRA in A-RZT roots than in leaves (**Figure 4D**) while NRA of 20°C-RZT leaves was much higher than that of roots (**Figure 4B**). However, this was not seen in Nai Bai plants (**Figure 3D**). High NO_3– assimilation rate in the root may be due to low reduced N

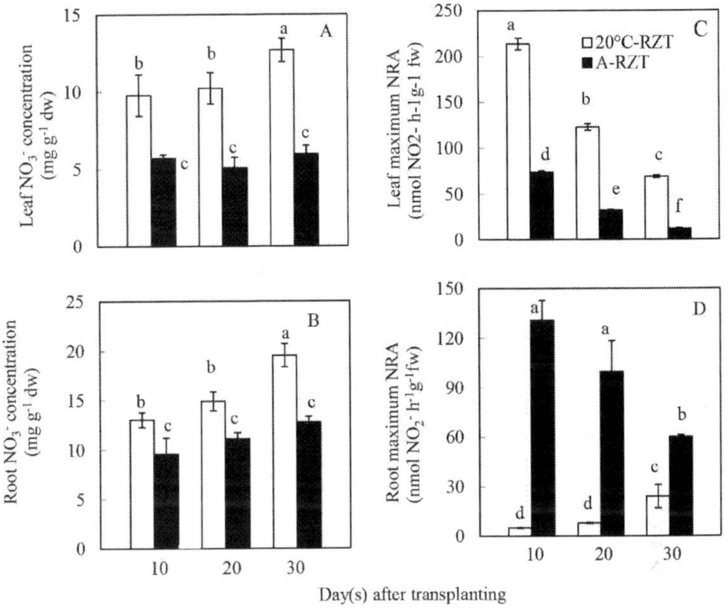

Figure 4. Leaf NO_3^- concentration (A) and maximum NRA (B), root NO_3^- concentration (C), and maximum NRA (D) of Baby butter head lettuce (*Lactuca sativa* L.) plant. Each point is the mean of five measurements of five different leaves from two different bins. Vertical bars represent the standard errors. Means with different letters above the columns are statistically different ($P < 0.001$) as determined by Tukey's multiple comparison test (unpublished data).

concentration in A-RZT roots (data not shown). Laurie and Stewart [60] grew cheakpea (*Cicer arietinum* L. ICARDA cultivar no. ILC 482) plants at high (40/25°C, day/night), and concentration in A-RZT roots, and moderate (25/17°C, day/night) temperature regimes. The results showed that high temperature (40/25°C) reduced shoot NRA (*in vivo*) but had little effect on root NRA [57].

On a total plant basis, high temperature growth shifted NRA from shoot to root, particularly in the young plant. Laurie and Stewart [60] also observed that there was a greater decline in leaf NRA with age [60]. This was in accordance with the present finding that leaf maximum NRA of 20°C-RZT Baby butterhead lettuce plants was highest in expanding young leaves (10 DAT) and lowest in fully matured leaves at 30 DAT (**Figure 4B**). Although leaf maximum NRA of A-RZT Baby butterhead lettuce plants decreased with the growth of plants, the decrease of NRA from 20 DAT to 30 DAT was not caused by the leaf age. Because the youngest fully expanded leaves were selected for the NRA analysis at both growth stages. Therefore, the repress of leaf maximum NRA in A-RZT plants at the late growth stage may be due to NR protein degradation caused by long-term supraoptimum RZT treatment. Although root maximum NRA was higher in A-RZT Baby butterhead lettuce plants compared to those of 20°C-RZT plants, the lower NO_3– uptake and transport capacity (lower root and leaf NO_3– concentration) still restricted NO_3– reduction of A-RZT plants. This was evidenced by the lower total reduced N concentration of leaf and root of A-RZT plants compared to 20°C-RZT plants, especially at 20 and 30 DAT (data not shown). For Nai Bai plants, NO_3^- reduction took place both in leaves and roots. However, the major proportion (more than 80%) of total plant NO_3^- reduction occurred in leaves for both 25°C-RZT and A-RZT plants (**Figure 3C** and **D**). At very early growth stage (10 DAT), supraoptimum A-RZT did not decrease maximum NRA in leaves and roots of Nai Bai plants. Because of very low growth rate of the plant at this very early growth stage, the lower NO_3^- content in A-RZT plants did not limit NO_3^- reduction. As a result, both of total reduced N concentrations in leaves and roots of A-RZT Nai Bai plants remained at the similar levels to those of 25°C-RZT plants at 10 DAT (data not shown). However, at 10 DAT, sugar concentrations (especially, glucose, fructose, and starch) both in leaves and roots of A-RZT Nai Bai plants increased significantly compared to 25°C-RZT plants (data not shown). Sugar accumulation in roots may be considered as a general consequence of impaired growth [61]. These indicate that, for Nai Bai plants, carbohydrate metabolism is more sensitive to supraoptimum RZT than NO_3^- metabolism. Controversial results of effect of high RZTs on N metabolism in shoots and roots have been reported [38, 39, 59–61]. In our study, it was confirmed that RZTs significantly affect N metabolism both in leaves and roots of subtropical vegetable Nai Bai (*B. chinensis* L (**Figure 3**) and temperate vegetable Baby butter head lettuce (**Figure 4**).

5. Conclusions

Our studies showed that high A-RZT inhibited root elongation, branching, and hair formation but increased root diameter of subtropical and temperate vegetables grown in the tropics. However, cooling the RZ promoted root growth and development as well as shoot productivities of aeroponically grown temperate and subtropical vegetables in the tropics. Manipulation of rhizosphere environment can alter not only root morphology but also physiological characteristics of roots to meet changes in shoot water and nutrient demand in response to atmospheric high temperature. Hot A-RZT caused shoot water deficit of aeroponically grown plants with continual spraying nutrient solution due to the negative water balance between water uptake by the root system and water loss from the shoot. Thus, subtropical and temperate vegetables grown in the tropics greenhouse had experienced mid-day and permanent water deficit when they were grown at A-RZT. However, water deficit of C-RZT plants was alleviated due to the larger root system. Compared to subtropical and temperate vegetables grown at hot A-RZT, cooling the RZ enhanced NO_3^- uptake and its assimilation of shoots. However, effects of RZT on NO_3^- assimilation of roots depend on species. Adequate levels of nutrient, especially N of C-RZT plants, alleviated nonstomatal limitation of photosynthesis.

Acknowledgements

All experiment work on subtropical and temperate vegetables grown aeroponically was supported by the Academic Research Fund (Grant reference numbers: RP10/95 HJ, RP 1/97 HJ,

RP 22/98 HJ, RP12/01, RI7/05 HJ, Ministry of Education, Singapore and Singapore Millennium Foundation (Project code: SMF-Farming System).

Author details

Jie He

Address all correspondence to: jie.he@nie.edu.sg

National Institute of Education, Nanyang Technological University, Singapore

References

[1] Malamy JE. Intrinsic and environmental response pathways that regulate root system architecture. Plant Cell, Environment. 2005; 28: 67–77, DOI: 10.1111/j.1365-3040.2005.01306.x/full.

[2] He J. Mineral nutrition of aeroponically grown subtropical and temperate crops in the tropics with manipulation of root-zone temperature at different growth irradiances. Plant Stress. 2010; 4: 14–30, http://www.globalsciencebooks.info/JournalsSup/10PS_4_SI2.html.

[3] He J. Impact of root-zone temperature on photosynthetic efficiency of aeroponically grown temperate and subtropical vegetable crops in the tropics. In: Buchner Th. B. Ewingen NH, Editors. Theory and Applications in Energy, Biotechnology and Nanotechnology. Nova Science Publishers Inc.: New York; 2009. pp. 111–143, http://trove.nla.gov.au/work/27991465?versionId=45343866. Ch 4.

[4] Hodge A, Berta G, Doussa C, Merchan F, Crespi M. Plant root growth, architecture and function. Plant Soil. 2009; 321: 153–187, DOI: 10.1007/s11104-009-9929-9.

[5] Kirk GJD, Van DL. Changes in rice root architecture, porosity, and oxygen and proton release under phosphorus deficiency. New Phytologist. 1997; 135: 191–200, DOI: 10.1046/j.1469-8137.1997.00640.x

[6] Neumann G, Massonneau A, Martinoia E, Römheld V. Physiological adaptations to phosphorus deficiency during proteoid root development in white lupin. Planta. 1999; 208: 373–382, http://link.springer.com/article/10.1007%2Fs004250050572.

[7] Xie Y, An S, Yao X, Xiao K, Zhang C. Short-time response in root morphology of *Vallisneria natans* to sediment type and water column nutrient. Aquatic Botany. 2005; 81: 85–96, DOI: 10.1016/j.aquabot.2004.12.001.

[8] Yang JC. Relationships of rice root morphology and physiology with the formation of grain yield and quality and the nutrient absorption and utilization. Scientia Agriculture Sinica. 2011; 44: 36–46, DOI: 10.3864/j.issn.0578-1752.2011.01.005.

[9] He J, Lee SK. Growth and photosynthetic characteristics of lettuce (*Lactuca sativa* L.) grown under fluctuating hot ambient temperatures with the manipulation of cool rootzone temperature. Journal of Plant Physiology. 1998; 152: 387–391, DOI: 10.1016/S0176-1617(98)80252-6.

[10] He J, Lee SK. Growth and photosynthetic responses of three aeroponically grown lettuce cultivars (*Lactuca sativa* L.) to different rootzone temperatures and growth irradiances under tropical aerial condition. Journal of Horticulture Sciences and Biotechnology. 1998; 73: 173–180, DOI: 10.1080/14620316.1998.11510961.

[11] Lynch J. Root architecture and plant productivity. Plant Physiology. 1995; 109: 7–13, DOI: 10.1104/pp.109.1.7.

[12] Tan LP, He J, Lee SK. Effects of root-zone temperature on the root development and nutrient uptake of *Lactuca sativa* L. cv 'Panama' grown in an aeroponic system in the tropics. Journal of Plant Nutrition. 2002; 25: 297–314, DOI: 10.1081/PLN-100108837.

[13] Qin L, He J, Lee SK, Dodd IC. An assessment of ethylene mediation of lettuce (*Lactuca sativa*) root growth at high temperatures. Journal of Experimental Botany. 2007; 58: 3017–3024, DOI: 10.1093/jxb/erm156.

[14] Kuroyanagi T, Paulsen GM. Mediation of high-temperature injury by roots and shoot during reproductive growth of wheat. Plant, Cell and Environment. 1988; 11: 517–523, DOI: 10.1111/j.1365-3040.1988.tb01790.x.

[15] Cumbu IP, Nye PH. Root zone temperature effects on growth and nitrate absorption in rape (*Brassica napus* cv. Emeralk). Journal of Experimental Botany. 1982; 33: 1138–1446, http://www.jstor.org/stable/23690485.

[16] Sattelmacher B, Marschner H, Kühne R. Effects of the temperature of the rooting zone on the growth and development of roots of potato (*Solanum tuberosum*). Annals Botany. 1990; 65: 27–36.

[17] Pardales JR, Yamauchi A, Kono Y. Growth and development of sorghum roots after exposure to different periods of a hot root-zone temperature. Environment Experimental Botany. 1991; 31: 397–403, DOI: 10.1016/0098-8472(91)90037-O.

[18] Menzel CM, Turner DW, Doogan VJ, Simpson PR. Root–shoot interactions in passionfruit (*Passiflora* sp.) under the influence of changing root volumes and soil temperatures. Journal of Horticulture Science. 1994; 69: 353–364, DOI: 10.1080/14620316.1994.11516487.

[19] Tan LP. Root-zone temperature effects on vegetable physiology in a commercial aeroponic system in Singapore. [thesis]. Nanyang Technological University, Singapore; 2004.

[20] Abbas Al-Ani MK, Hay RKM. The influence of growing temperature on the growth and morphology of cereal seedling root systems. Journal of Experimental Botany. 1983; 34: 1720–1730, DOI: 10.1093/jxb/34.12.1720.

[21] Pierik R, Verkerke W, Voesenek RACJ, Blom KWPM, Visser EJW. Thick root syndrome in cucumber (*Cucumis sativus* L.): a description of the phenomenon and an investigation of the role of ethylene. Annals Botany. 1999; 84: 755–762, DOI: 10.1006/anbo.1999.0980.

[22] Arshad M, Frakenberger WT Jr. Ethylene: Agricultural Sources and Applications. Kluwer Academic/Plenum publishers: New York. 2002; pp. 155–156.

[23] Choong TW, He J, I. Dodd. Growing *Lactuca* in an aeroponic system within a tropical greenhouse – lower accumulated rootzone ethylene concentrations corresponded with higher shoot growth at cooler rootzone temperatures. *Frontier* (accepted).

[24] He J, Tan LP, Lee SK. Root-zone temperature effects on photosynthesis, ^{14}C-photo-assimilate partitioning and growth of temperate lettuce (*Lactuca sativa* cv. 'Panama') grown in the tropics. Photosynthetica. 2009; 47: 95–103, DOI: 10.1007/s11099-009-0015-6.

[25] Minchin PEH. Thorpe, M.R. What determines carbon partitioning between competing sinks? Journal of Experimental Botany. 1996; 47: 1293–1296, DOI: 10.1093/jxb/47.

[26] Minchin PEH, Farrar JF, Thorpe MR. Partitioning of carbon in split root system of barley: effect of temperature of the root. Journal of Experimental Botany. 1994; 45: 1103–1109, DOI: 10.1093/jxb/45.8.1103.

[27] Schurr U, Heckenberger U, Herdel K, Walter K, Feil R. Leaf development in *Ricinus communis* during drought stress: dynamics of growth processes, of cellular structure and of sink-source transition. Journal of Experimental Botany. 2000; 51: 1515–1529, DOI: 10.1093/jexbot/51.350.1515.

[28] Klock KA, Taber HG, Graves WR. Root respiration and phosphorus nutrition of tomato plants grown at a 36C root-zone temperature. Journal of the American Society for Horticultural Science. 1997; 122: 175–178, http://cat.inist.fr/?aModele= afficheN&cpsidt=10736550.

[29] Forde BG, Clarkson DT. Nitrate and ammonium nutrition of plants: Physiological and molecular perspectives. Advances in Botanical Research. 1999; 30: 1–90, DOI: 10.1016/S0065-2296(08)60226-8.

[30] McCully M. How do real roots work? Some new view of root structure. Plant Physiology. 1985; 109: 1–6, DOI: 10.1104/pp.109.1.1.

[31] Grime JP, Crick JC, Rincon JE. The ecological significance of plasticity. Symposia of the Society for Experimental Biology. 1986; 40: 5–29, http://www.ncbi.nlm.nih.gov/pubmed/3544310.

[32] Cooper AJ. Root temperature and plant growth: a review. In: Commonwealth Bureau of Horticulture and Plantation Crops. Commonwealth Agriculture Bureaux, Farnham Royal, Slough: England, 1973. p. 73, http://trove.nla.gov.au/version/25843400.

[33] Delucia EH, Scott AH, Day TA. Effects of soil temperature on growth, biomass allocation and resource acquisition of *Andropogon gerardii* Vatman. New Phytologist. 1992; 10: 543–549, http://www.jstor.org/stable/2557416.

[34] He J, Chua NYA, Qin, L. Interaction between iron stress and root-zone temperature on physiological aspects of aeroponically grown Chinese broccoli (*Brassica alboglabra*). Journal of Plant Nutrition. 2008; 31: 1–20, DOI: 10.1080/01904160701742121.

[35] Luo HY. Mechanisms of Ca, P and K uptake and partitioning characteristics of certain temperate vegetables in relation to different irradiance and root-zone temperatures [thesis]. Nanyang Technological University, Singapore; 2008.

[36] Engels C, Münkle L, Marschner H. Effect of root zone temperature and shoot demand on uptake and xylem transport of macronutrients in maize (*Zea mays* L.). Journal of Experimental Botany. 1992; 43: 537–547, DOI: 10.1093/jxb/43.4.537.

[37] Macay AD, Barber SA. Soil temperature on root growth and phosphorus uptake by corn. Soil Science Society of America. 1984; 48: 818–823, DOI: 10.2136/sssaj1984.03615995004800040024x.

[38] He J, Lee SK. Relationship among photosynthesis, Ribulose-1,5-Bisphosphate Carboxylase (Rubisco) and water relations of subtropical vegetable Chinese broccoli grown in the tropics by manipulation of root-zone temperature. Environmental and Experimental Botany. 2001; 46: 119–128, DOI: 10.1016/S0098-8472(01)00089-2.

[39] He J, Lee SK, Dodd IC. Limitations to photosynthesis of lettuce grown under tropical conditions: alleviation by root-zone cooling. Journal of Experimental Botany. 2001; 52: 1323–1330, DOI: 10.1093/jexbot/52.359.1323.

[40] Turner NC. Stomatal behavior and water status of maize, sorghum, and tobacco under field conditions II. At low soil water potential. Plant Physiology. 1974; 53: 360–365, http://www.ncbi.nlm.nih.gov/pmc/articles/PMC367352/.

[41] Turner NC, Jones MM. Turgor maintenance by osmotic adjustment. In: Turner NC. Kramer PJ. Editors. A Review and Evaluation. Adaptation of Plants to Water and High Temperature Stress. Wiley-Interscience: New York, USA; 1980. pp. 87–103, http://www.cabdirect.org/search.html?q=turgor+maintenance+by+osmotic+adjustment.+a+review+and+evaluate.

[42] Dodd IC, He J, Turnbull, CGN, Lee SK, Critchley C. Influence of supra-optimal root temperatures on growth and stomatal conductance in *Capsicum annuum* L. Journal of Experimental Botany. 2000; 51: 239–248, DOI: 10.1093/jexbot/51.343.239.

[43] Marschner H. Mineral Nutrition of Higher Plants. Academic: London, UK ; 1995. p. 889, http://www.sciencedirect.com/science/book/9780124735422

[44] der Leij M, Smithe SJ, Miller AJ. Remobilization of vacuolar stored nitrate in barley root cells. Planta. 1998; 205: 64–72, http://link.springer.com/article/10.1007/s004250050297.

[45] Atkin OK, Cummins WR. The effect of root temperature in the induction of nitrate reductase activities and nitrogen uptake rates in arctic plant species. Plant and Soil. 1994; 159: 187–197, DOI: 10.1007/BF00009280.

[46] Engels C. Effect of root and shoot meristem temperature on shoot and root dry mater partitioning and internal concentrations of nitrogen and carbohydrates in maize and wheat. Annals Botany. 1994; 73: 211–219, DOI: 10.1006/anbo.1994.1025.

[47] Ali IA, Kafkafi U, Yamaguchi I, Sugimoto Y, Inanaga S. Response of oilseed rape plant to low root temperature and nitrate:ammonium ratios. Journal of Plant Nutrition. 1998; 21: 1463–1481, DOI: 10.1080/01904169809365496.

[48] Du YC, Tachibana S. Effect of supraoptimal root temperature on the growth, root respiration and sugar content of cucumber plants. Scientia Horticulturae. 1994; 58: 289–330, DOI: 10.1016/0304-4238(94)90099-X.

[49] Yeager TH, Harrison RH, Ingram DL. 'Rotundifolia' holly growth and nitrogen accumulation influenced by supraoptimal root-zone temperatures. HortScience. 1991; 26: 1381–1388, http://hortsci.ashspublications.org/content/26/11/1387.full.pdf.

[50] Johnson CR, Ingram DL. *Pittosporum tobira* response to container medium temperature. HortScience. 1984; 19: 524–525, http://openagricola.nal.usda.gov/Record/GUA84131934.

[51] Gosselin A, Trudel MJ. Interaction between root-zone temperature and light levels on growth, development and photosynthesis of *Lycopersicon esculentum* Mill. culticar 'vendor'. Scientia Horticulturae. 1984; 23: 313–321. www.sciencedirect.com/science/article/pii/030442388490027X

[52] Cruz C, Lips SH, Martinsloucao MA. Effect of root temperature on carob growth – nitrate versus ammonium nutrition. Journal of Plant Nutrition. 1993, 16: 1517–1530, DOI: 10.1080/01904169309364629.

[53] Atkinson CJ, Ruiz LP, Mansfield TA. Calcium in xylem sap and the regulation of its delivery to the shoot. Journal of Experiment Botany. 1992; 43: 1315–1324, DOI: 10.1093/jxb/43.10.1315.

[54] Solomonson LP, Barber MJ. Assimilatory nitrate reductase: functional properties and regulation. Annual Review of Plant Physiology and Plant Molecular Biology. 1990; 41: 225–253, DOI: 10.1146/annurev.pp.41.060190.001301.

[55] Cheng CL, Dewdney J, Kleinhofs A, Goodman HM. Cloning and nitrate induction of nitrate reductase mRNA. Proceedings of the National Academy of Sciences USA. 1986; 83: 6825–6826, http://www.ncbi.nlm.nih.gov/pmc/articles/PMC386602.

[56] Savidov NA, Tokarev BI, Lips SH. Regulation of Mo-cofactor, NADH- and NAD(P)H-specific nitrate reductase activities in the wild type and two *nar*-mutant lines of barley

(*Hordeum vulgare* L.). Journal of Experiment Botany. 1997; 48: 847–855. DOI: 10.1093/jxb/48.4.847.

[57] Cen YP, Layzell DB. *In vivo* gas exchange measurement of the site and dynamics of nitrate reduction in soybean. Plant Physiology 2003; 131: 1147–1156, http://www.plantphysiol.org/content/131/3/1147.full.pdf.

[58] Scheurwater I, Koren M, Lambers H, Atkin OK. The contribution of roots and shoots to whole plant nitrate reduction in fast- and slow-growing grass species. Journal of Experimental Botany. 2002; 55: 1635–1642.

[59] Gojon A, Passard C, Bussi C. Root/shoot distribution of NO_3^- assimilation in herbaceous and woody species. In: Roy R, Garnier E, editors. A Whole Plant Perspective on Carbon-Nitrogen Interactions. SPB Academic Publishing: The Hague; 1994. pp. 131–147, http://link.springer.com/article/10.1023/A%3A1000903504144

[60] Laurie S, Stewart GR. Effects of nitrogen supply and high temperature on the growth and physiology of the chickpea. Plant, Cell and Environment. 1993; 16: 609–621, DOI: 10.1111/j.1365-3040.1993.tb00479.

[61] Hansch R, Fessel DG, Witt C, Hesberg C, Hoffmann G, Walch-Liu P, Engels C, Kruse J, Rennenberg H, Kaiser W, Mendel RR. Tobacco plants that lack expression of functional nitrate reductase in roots show changes in growth rates and metabolite accumulation. Journal of Experimental Botany. 2001; 52: 1251–1258, DOI: 10.1093/jexbot/52.359.1251.

Chapter 10

Molecular and Morphophysiological Analysis of Drought Stress in Plants

Summy Yadav and Kamal Dutt Sharma

Additional information is available at the end of the chapter

http://dx.doi.org/10.5772/65246

Abstract

Drought is a major environmental stress factor that affects the growth and development of plants. Most of the physiological traits associated with drought tolerance are quantitative in nature. An important research strategy that has been widely used to deal with such complexity is to use molecular markers to identify quantitative trait loci (QTLs) in appropriate mapping populations. In response to drought brought about by soil water deficit, plants can exhibit either drought escape or drought resistance mechanisms, with resistance further classified into drought avoidance and drought tolerance. Drought escape is the ability of plants to complete the life cycle before severe stress arrives. Drought avoidance is the maintenance of high tissue water potential in spite of soil water deficit. Drought avoidance is consequence of improved water uptake under stress and the capacity of plant cells to hold acquired water that reduces water loss. Drought tolerance is the ability to withstand water deficit with low tissue water potential. Plant water status that includes leaf water potential, osmotic potential and relative water content (RWC) represents an easy measure of water deficit and provides best sensor for stress. Genomics-assisted breeding (GAB) approaches, such as marker-assisted selection (MAS), can greatly improve precision and efficiency of selection in crop breeding. Molecular markers can facilitate indirect selection for traits that are difficult or inconvenient to score directly, pyramiding genes from different sources and combining resistance to multiple stresses. Conventional breeding for developing drought-tolerant crop varieties is time-consuming and labor intensive due to the quantitative nature of drought tolerance and difficulties in selection for drought tolerance. The identification of genomic regions associated with drought tolerance would enable breeders to develop improved cultivars with increased drought tolerance using marker-assisted selection (MAS). This requires integration of knowledge from plant physiology and biotechnology into plant breeding. The availability of a large number of molecular markers, dense genetic maps and markers associated with traits and transcriptomics resources have made it possible to integrate genomics technologies into chickpea improvement.

Keywords: abiotic stress, drought, physiological traits, conventional breeding, marker-assisted selection, QTLs

1. Introduction

Plant growth and productivity is adversely affected by nature's wrath in the form of various biotic and abiotic stress factors. Water deficit is one of the major abiotic stresses, which adversely affects crop growth and yield. Stress is an altered physiological condition caused by factors that tend to disrupt the equilibrium. Strain is any physical and chemical change produced by a stress [1]. Stress is used with various meanings, the physiological definition and appropriate term as responses in different environmental situations. If a factor deviates from its optimum does not necessarily results in stress. Stress is a constraint or unpredictable change imposed on regular metabolic patterns of growth results in injury, disease or aberrant physiology. Plants are mainly exposed to stresses such as drought, precipitation, salt, flooding, heat, oxidative stress and heavy metal toxicity. Drought stress occurs when the available water in the soil is reduced and atmospheric conditions cause continuous loss of water by transpiration or evaporation due to increase in temperature in nature. Drought stress tolerance is seen in almost all plants but its extent varies from species to species and even within species [2]. Conventional plant breeding attempts have changed over to use physiological selection criteria since they are time consuming and rely on present genetic variability [3]. Abiotic stresses tolerance is a complex trait, due to the interactions between stress factors and various molecular, biochemical and physiological phenomena affecting plant growth at different developmental stages [4]. High yield potential under drought stress is the target of crop breeders. In many cases, high yield potential can contribute to yield in moderate stress environment [5]. Drought stress leads to stomatal closure and limitation of gas exchange. Desiccation is much more extensive loss of water, which can potentially lead to maximum disruption of metabolism and cell structure and finally stops enzyme catalyzed reactions [6, 7]. Drought stress is characterized by reduction in water content, diminished leaf water potential and turgor loss, closure of stomata and decrease in cell enlargement and growth. Severe water stress may result in hampering photosynthesis, disturbing the overall metabolism and finally the necrosis of plant [8]. Water stress inhibits cell enlargement more as compared to cell division. Plant growth is reduced by affecting various physiological and biochemical processes, such as photosynthesis, respiration, translocation, ion uptake, carbohydrates, nutrient metabolism and growth promoters [9]. A better understanding of the morphophysiological traits can be used to create new varieties of crops to obtain a better productivity under drought conditions [10]. The reactions of plants to water stress differ significantly at various organizational levels depending upon intensity and duration of stress as well as plant species and its stage of growth [11]. A fundamental part for making the crops stress tolerant is to understand plant responses to different drought stress environments [12].

In response to drought brought about by soil water deficit, plants can exhibit either drought escape or drought resistance mechanisms, with resistance further classified into drought avoidance (maintenance of tissue water potential) and drought tolerance [13]. Drought stress is the ability of plants to complete the life cycle before severe stress conditions arise. Drought avoidance is the maintenance of high tissue water potential under a soil water deficit. Improved water uptake under stress and the capacity of plant cells to hold acquired water reduces water loss leading to drought avoidance. Drought tolerance is the ability to withstand water deficit with low tissue water potential. Plants respond to water deficit using mechanisms of avoidance by improved root traits and by reducing water loss through reduced epidermal (stomatal and cuticular) conductance, reduced radiation absorption, and reduced evaporative surface (leaf area). Drought tolerance is the ability to withstand water deficit with low tissue water potential [14]. Plants under drought stress may survive by, among other mechanisms, maintaining cell turgor and reducing evaporative water loss by accumulating compatible solutes [15]. In recent years, much molecular information has been generated on the response of plants to environmental stresses. Plants respond to environmental stresses such as drought by the induction of both regulatory and functional sets of genes [16, 17]. Very little is known about the early events in the perception of stress signals [18, 19]. The common stress signaling pathways have been distinguished into abscisic acid (ABA) dependent and ABA independent [20, 21]. Most of the key genes in these pathways have been identified, such as transcription factors belonging to the class of dehydration responsive element-binding protein (DREB)/C-repeat-binding factor (CBF), ABA-binding factor (ABF), Myelocytomatosis oncogene (MYC) and Myeloblastosis oncogene (MYB), including the identification of the stress-responsive cis-elements ABA-responsive element (ABRE) and dehydration responsive element (DRE). Downstream of the early signal perception events, signaling genes and molecules acting as secondary messengers have been identified, revealing the role of Ca^{+} and reactive oxygen species (ROS) as secondary messengers. These regulatory mechanisms induce downstream functional genes, which are needed to establish new cellular homeostasis that leads to drought tolerance and/or resistance.

Most of the physiological traits associated with drought tolerance are quantitative in nature. Genomics-assisted breeding (GAB) approaches, such as marker-assisted selection (MAS), can greatly improve precision and efficiency of selection in crop breeding [23]. Integration of genomics and breeding has a great potential for crop improvement. Molecular markers facilitate indirect selection for traits that are inconvenient to score directly (e.g., root traits, resistance to root knot nematodes), pyramiding genes from different sources (e.g., bringing together ascochyta blight resistance genes from different donors) and combining resistance to multiple stresses (e.g., resistance to fusarium wilt and ascochyta blight). Recent years have seen tremendous progress in the development of large scale genomic resources such as DNA-based molecular markers, comprehensive genetic maps, whole-genome transcription profiling techniques to identify genomic regions and genes underlying plant stress responses [24]. These genomic tools will be useful to understand and access the diversity conserved in ex situ germplasm collections for crop improvement [25]. Thus, an understanding of drought stress and water use in relation to plant growth is of importance for sustainable agriculture. Conventional breeding for developing drought-tolerant crop varieties is time-consuming and labor intensive due to the quantitative nature of drought tolerance and difficulties in selection for drought tolerance [26]. Mapping of different genomes has been of interest to identify genomic locations of disease resistance genes and other yield-related traits. Isolation and validation of genes underlying the QTL/genes for the traits of interest is an essential step to determine gene function. QTLs for drought tolerance have been identified for major important crop species such as rice, maize, wheat, barley, sorghum, pearl millet, soybean and chickpea. These QTLs were identified for many important traits which include yield and yield-related traits under drought stress conditions, physiological responses including water-soluble carbohydrates, carbon isotope ratio, osmotic potential, chlorophyll content, flag leaf rolling index, grain carbon isotope discrimination, relative water content, leaf osmotic potential, osmotic adjustment, chlorophyll and chlorophyll fluorescence parameters to drought stress, flowering time, root traits. Major QTLs contribute to the traits with higher phenotypic variation. These QTLs, after validation in desired germplasm, can be used for introgressing drought tolerance from the donor genotypes into less drought-tolerant cultivars or breeding lines (recipient parents) avoiding transfer of undesirable or deleterious genes from the donors (linkage drag).

2. Drought stress improvement

Drought can be defined as below normal precipitation that limits plant productivity. Drought can be classified as either terminal or intermittent. The availability of soil water decreases progressively during terminal drought, and it may lead to severe drought stress at the later period of crop growth and development. Finite periods of inadequate rain or irrigation occurring at one or more intervals during the growing seasons is the condition of intermittent drought [27]. According to Crosser [28] drought delays formation of sugars, lowers energy exchange and destroys the entire biochemical processes. Heat stress at sowing directly affects crop germination and crop establishment. Chickpea seed germination decreases at supra-optimum temperatures [29]. Ellis *et al.* [30] indicated that the optimal temperature for germination is 10–15°C and noted that high germination temperatures are considered to be 22–35°C. The adaptive strategies to high temperature stress are classified into the following three groups [31].

2.1. Drought escape

Drought escape can be defined as the ability of a plant to complete its life cycle before a serious plant water deficit develops. Plants can escape heat stress with early phenological development (early flowering and early maturity), developmental plasticity (variation in duration of growth period depending on the extent of water-deficit) and remobilization of pre-anthesis assimilates to grain [32]. Though flower initiation is sensitive to rising temperature in chickpea [33], early flowering and maturity is a heat escape mechanism [34] particularly in the Mediterranean spring-sown environments and south Indian germplasm. Flowering time is an important trait related to drought adaptation, where a short life cycle can lead to drought escape [35]. Crop duration is interactively determined by genotype and the environment and determines the ability of the crop to escape from climatic stresses including drought. Matching growth duration of plants to soil moisture availability is critical to realize high seed yield [36]. Drought escape occurs when phenological development is successfully matched with periods of soil moisture availability, where the growing season is shorter and terminal drought stress predominates. In field-grown clones of Robusta coffee, leaf shedding in response to drought stress revealed that drought-sensitive clone has greater extent of leaf shedding [37]. Time of flowering is a major trait of a crop adaptation to the terminal drought and high temperature environments. Short-duration varieties to be developed to minimize yield loss from terminal drought, as early maturity helps the crop to avoid the period of stress [38]. However, yield is generally correlated with the length of crop duration under favorable growing conditions, and any decline in crop duration below the optimum would tax yield [39].

2.2. Drought avoidance

The ability of plants to maintain relatively high tissue water potential despite a shortage of soil-moisture; mechanisms for improving water uptake, storing in plant cell and reducing water loss confer drought avoidance is referred to as drought avoidance. different mechanisms for drought avoidance are being reported in different plant species which include maintenance of turgor through increased rooting depth, efficient root system and increased hydraulic conductance and reduction in water loss through reduced epidermal (stomatal and lenticular) conductance, reduced absorption of radiation by leaf rolling or folding and reduced evaporation surface (leaf area). In crops, high root biomass has been of interest because the more the roots, the more their efficiency in absorption of water. This gives a plant more advantage in times when less moisture is available in the soil. A positive correlation between root system sizes and resistance to water stress has been found in several crops and many breeding attempts have focused on obtaining cultivars with larger root systems [40]. Saxena [41] has developed a chickpea cultivar with a greater degree of drought tolerance from combining large root traits of ICC4958. Similarly, Krishnamurthy et al. [42] has reported that large root biomass in a minicore collection of ICRISAT chickpea germplasm had high correlation with drought tolerance. Root system size is a complex trait since it is determined by intrinsic genetic factors and modulated by numerous environmental cues such as nutrient and moisture availability in the soil [43]. He also noted that smaller leaf surface was also a desirable trait related to drought tolerance. Plants with small leaf surface (pinnules) have shown to experience reduced water loss [41]. Glaucousness or waxy bloom on leaves helps with maintenance of high tissue water potential and is therefore considered as a desirable trait for drought tolerance [44, 45]. Varying percentage of glaucousness in wheat led to increased water-use efficiency, but it has minimal affect on total water use or harvest index. Determination of leaf temperature indicated glaucous leaves were 0.7°C cooler than non-glaucous leaves and had a lower rate of leaf senescence [43]. It was also suggested that a 0.5°C reduction in leaf temperature for 6 h per day was sufficient to extend the grain-filling period by more than three days. However, yield advantages are likely to be small as many varieties already show some degree of glaucousness.

2.3. Drought tolerance

The ability of plant to withstand water-deficit with low tissue water potential is referred as drought tolerance. A balance between maintenance of turgor and reduction in water loss helps plants to survive drought stress conditions [46]. Plants can combat drought stress by maintenance of turgor through osmotic adjustment (a process which induces solute accumulation in cell), increase in elasticity in cell and decrease in cell size and desiccation tolerance by protoplasmic resistance [47]. Drought resistance is increased by maintaining plant turgor pressure. Drought tolerance characters studied are primarily involved with protection of cellular structure from the effect of cellular dehydration. Dehydrins and late-embryogenesis abundant (LEA) proteins are being accumulated in response to decrease in plant tissue water content [48]. These proteins are said to act as chaperones that protect protein and membrane structure [49]. Compatible solutes can also protect protein and membrane structure under dehydration [50]. The role of reactive oxygen species (ROS) in stress signaling have been extensively studied in recent years and reviewed [51, 52]. Consequently, crop adaptation must reflect a balance among escape, avoidance and tolerance while maintaining adequate productivity. Use of these traits as indirect selection for grain yield has been reported to be easier in breeding programs than selection based on direct grain yields [53].

2.4. Antioxidant defense

The antioxidant defense system in the plant cell constitutes both enzymatic and non-enzymatic components. Enzymatic components include superoxide dismutase, catalase, peroxidase, ascorbate peroxidase and glutathione reductase. Non-enzymatic components contain cysteine, reduced glutathione and ascorbic acid [54]. High activities of antioxidant enzymes and high contents of non-enzymatic constituents are important under drought stress conditions.

The reactive oxygen species in plants are removed by a variety of antioxidant enzymes and/or lipid-soluble and water soluble scavenging molecules [55] the antioxidant enzymes being the most efficient mechanisms against oxidative stress [56]. Along with catalase, various peroxidases and peroxiredoxins enzymes are involved in the ascorbate-glutathione cycle, this pathway allows the scavenging of superoxide radicals and H_2O_2. Different enzymes that metabolize glutathione cycle are ascorbate peroxidase, dehydroascorbate reductase, monodehydroascorbate reductase and glutathione reductase [57]. Most of the glutathione cycle enzymes are found in the cytosol, stroma of chloroplasts, mitochondria and peroxisomes. Ascorbate peroxidase is a key antioxidant enzyme in plants whilst glutathione reductase has a central role in maintaining the reduced glutathione pool during stress [58]. Two glutathione reductase complementary deoxyribonucleic acids have been isolated; one type encoding the cytosolic isoforms and the other encoding glutathione reductase proteins dual-targeted to both chloroplasts and mitochondria in different plants [59].

Superoxide dismutase plays an important role, it catalyzes the dissociation of two molecules of superoxide into O_2 and H_2O_2. Lima et al. [60] proposed that drought tolerance of a particular plant species can be associated with enhanced activity of antioxidant enzymes. In contrast, Pinheiro et al. [61] in his studies on four clones of *Coffea canephora* did not find a link between protection against oxidative stress and drought tolerance. Oxidative damage in the plant tissue is alleviated by both enzymatic and non-enzymatic antioxidant systems. These include β-carotenes, ascorbic acid, α-tocopherol, reduced glutathione and enzymes including superoxide dismutase, peroxidase, ascorbate peroxidase, catalase, polyphenol oxidase and glutathione reductase [62, 63]. Carotenes are crucial part of the plant antioxidant defense system [64]; in

http://dx.doi.org/10.5772/65246

spite of this, they are very susceptible to oxidative destruction. The β-carotene present in the chloroplasts of all green plants is exclusively bound to the core complexes of photosystem I and photosystem II. Protection against damaging effects of reactive oxygen species at this site is essential for chloroplast functioning. β-carotene functions as an accessory pigment, it also acts as an effective antioxidant and plays a unique role in protecting photochemical processes and sustaining them. β-carotene also has a protective role in photosynthetic tissue by direct quenching of triplet chlorophyll, which prevents the generation of singlet oxygen and protects from oxidative damage.

2.5. Plant growth regulators

Plant growth regulators phytohormones are substances that influence physiological processes of plants at very low concentrations, either they are applied externally or produced in the plant [65]. Both these terms have been used interchangeably, particularly when referring to auxins, gibberellins, cytokinins, ethylene and abscisic acid [66]. Under drought, endogenous contents of auxins, gibberellins and cytokinin usually decrease, while those of abscisic acid and ethylene increase [67]. Nevertheless, phytohormones play vital roles in drought tolerance of plants. Auxins break root apical dominance helping in new root formation induced by cytokinins. Drought stress limits the production of endogenous auxins, usually when contents of abscisic acid and ethylene increase. Application of indole-3-yl-acetic acid exogenously enhanced net photosynthesis and stomatal conductance in cotton (Kumar et al., 2001). Indole-3-butyric acid is a naturally occurring auxin. Enhanced indole-3-butyric acid synthesis was observed in maize in response to drought stress and abscisic acid application. Enzyme indole-3-butyric acid synthetase was revealed from *Arabidopsis* under drought stress [68]. Experiments with indole-3-yl-acetic acid and ethylene glycol tetra-acetic acid suggested that calcium and auxin participate in signaling mechanisms of drought-induced proline accumulation [69]. An adaptive strategy that occurs during progressive drought stress is drought rhizogenesis. Families such as Brassicaceae form short and tuberized, hairless roots in response to drought stress. These roots are capable of withstanding a prolonged drought period and give rise to a new functional root system upon rehydration. The drought rhizogenesis was highly increased in the gibberrelic acid biosynthetic mutant *ga5*, suggested that gibberrelic acids also participate in this process [70]. Abscisic acid is a growth inhibitor and produced under a wide variety of environmental stresses. All plants respond to drought and many other stresses by accumulating abscisic acid. Abscisic acid is ubiquitous in all flowering plants and is generally recognized as a stress hormone that regulates gene expression and acts as a signal for the initiation of processes involved in adaptation to drought and other environmental stresses. It has been proposed that abscisic acid and cytokinin have opposite roles in drought stress. Increase in abscisic acid and decline in cytokinins levels favor stomatal closure and limit water loss through transpiration under water stress [71]. Increased abscisic acid concentration leads to many changes in development, physiology and growth. Abscisic acid alters the relative growth rates of various plant parts such as increase in the root-to-shoot dry weight ratio, inhibition of leaf area development and production of prolific and deeper roots. It triggers the occurrence of a complex series of events leading to stomatal closure, which is an important water conservation response [72]. In a study on genetic variation for abscisic acid accumulation in rice, a consistent negative relationship between the ability of detached and partially dehydrated leaves to accumulate abscisic acid and leaf weight was established [73]. By its effect in closing stomata, abscisic acid can control the rate of transpiration and, to some extent, may be involved in the mechanism conferring drought tolerance in plants.

Ethylene is considered as growth inhibitory hormone, it is involved in environmentally driven growth inhibition and stimulation [66]. The response of cereals to drought includes loss of leaf function and premature onset of senescence in older leaves. Ethylene regulates leaf performance throughout its lifespan as well as to determine the onset of natural senescence and mediate drought-induced senescence [74]. Recent studies suggest that growth promotion is a common feature in ethylene responses. To escape this adversity, plants can optimize growth and tolerate abiotic stresses such as drought, and this response also involves ethylene synthesis [75].

Polyamines are known to have profound influence on plant growth and development. Being cationic, polyamines can associate with anionic components of the membrane, such as phospholipids, thereby protecting the lipid bilayer from deteriorating effects of stress. There has been a growing interest in the study of polyamine participation in the defense reaction of plants against environmental stresses and extensive research efforts have been made in the last two decades [76, 77]. Different genes for enzymes involved in polyamine metabolism has been analyzed for their expression under drought stress in several species. For example, the apple spermidine synthase gene when overexpressed encodes high levels of spermidine synthase, which substantially improves abiotic stress tolerance including drought [78].

3. Morphophysiological mechanisms for drought stress in plants

Water limitation is one of the important factors limiting crop productivity worldwide. Nearly all terrestrial plants are exposed to drought stress at different times and to different intensities during their life cycle [79, 80]. As water is fundamental to almost all aspects of plant growth, plants are thought to have evolved numerous strategies for coping with limited water availability including changes in phenological developmental and physiological traits [81, 82].

3.1. Phenological traits

3.1.1. Early flowering and maturity

Early maturity is an important trait to avoid drought stress. Early flowering and early podding are two main components of drought escape in crops to avoid higher yield losses from drought. The differential genotypic response to drought stress, as a result of variation in physiological parameters has also been reported by Gunes et al. [83]. Early maturing chickpea varieties that escape terminal drought have been developed, but early maturity decreases yield and limits the crop's ability for extended growing periods. Chickpea genotypes with high growth vigor showed early maturity. Selection for high growth vigor enhances chances for escaping terminal drought stress [84]. Initial growth vigor is suitable character for large-scale evaluation of germplasm and breeding materials [85].

3.1.2. Root and shoot traits

Extensive and deep root systems have been recognized as one of the most important traits for improving crop productivity under progressively receding soil moisture condition. Roots have a major role in dehydration avoidance as deep root system is able to obtain moisture from the deeper soil layers even when the upper soil layer becomes dry. The root traits such as biomass, length density and depths have been proposed as the main drought avoidance traits to contribute to seed yield under terminal drought environment [83]. Upadhyaya et al. [86] observed chickpea variety ICC13124 was equally good in respect of root traits (root length, root weight and root volume) as compared to ICC4958. Shoot fresh weights were significantly greater in well watered genotypes, but there was no significant effect of moisture stress on shoot dry matter content, revealing that weight of fresh shoot was higher due to high uptake of water under well watered conditions which evaporated after drying.

3.2. Physiological traits

3.2.1. Leaf water status

Moisture deficit affects plant establishment in the field, photosynthetic ability and osmotic behavior of cells. However, species and genotypes vary in their capacity to tolerate water stress [87]. Plants adopt various defense mechanisms in response to terminal drought which are accomplished by regulating internal plant water status. Plant water status that includes leaf water potential, osmotic potential and relative water content represents an easy measure of water deficit and provides best sensor for stress. Water stress reduces the osmotic potential of tissues in the plant which helps in maintenance of turgor potential for normal metabolic activities which has been recognized as basic mechanism of drought tolerance [88]. Gupta et al. [89] studied the physiological mechanism of drought tolerance in chickpea. It was observed that tolerant genotype had lower membrane injury, retain imbibitions seedling growth, osmotic adjustment and water use efficiency. A partial closer of stomata led to decreased conductance under water stress resulting into reduced transpiration and photosynthesis has been reported by (Sharma and Singh) [90]. Kushwaha et al. [91] indicated that genotypes which possessed high initial water content (IWC) along with high relative water content resulted in relatively less damage to the assimilatory system resulted in to the production of relatively higher biomass. The osmo-regulatory activities helped the plant to cope up with moisture stress. Variation in RWC is achieved through differences in plant ability to absorb water from soil by developing a high water potential gradient from soil to plant, extending rooting depth or ability to control water loss through stomata [92]. A decrease in the relative water content (RWC) in response to drought stress has been recorded in wide variety of plants as reported by Nayyar and Gupta [93].

3.2.2. Relative stress injury, CTD and photochemical efficiency

The role of cell membrane remains to be more critical for adaptation under temperature and moisture stress conditions. Blum and Ebercon [94] described that under water stress conditions measurement of electrolyte leakage can be used to estimate water stress tolerance. Heat tolerant genotypes were able to posses higher membrane stability [95]. Higher membrane stability in drought tolerant genotypes under stress was due to increased activities of antioxidative enzymes which prevent damage of membrane by active oxygen species produced under stress. It had been reported that tolerant and intermediate genotypes were superior to susceptible ones in maintaining membrane stability and lower membrane injury under drought stress condition [96]. Stomatal closure occurs when plants are subjected to water stress in order to decrease energy dissipation. Transpiration plays a major role in leaf cooling and reduces canopy temperature relative to ambient temperature. Relatively lower canopy temperature in drought stressed crop plants indicates a relatively better capacity for taking up soil moisture and for maintaining a relatively better plant water status. The photosynthetic efficiency, transpiration and the values of relative stress injury declined in chickpea under drought conditions [97]. Photosynthetic pigments play an important role in light harvesting and dissipation of excess energy. It is known that the content of both chlorophyll a and b changes under drought stress [98]. Carotenoids participate in energy dissipation and can aid plant resistance against drought stress.

4. Breeding for drought tolerance

Drought offers great challenges to plant breeders around the globe. Drought is usually uncertain and unpredictable in the field and response of canopy toward drought is perceived using conventional techniques mainly. Conventional breeding procedures such as introduction, selection, hybridization and mutation are widely used by breeders. In spite of conventional methods novel methods such as *in situ* and *in vitro* techniques can also **be** used for selection, survival rate or to monitor gene expression changes of wild-type plants genotypes overexpressing candidate genes for drought tolerance. Plant responses to drought at both the physiological and molecular levels are studied extensively. Major drawback of studies for drought treatments is uncontrolled soil water moisture and comparison of performance of different genotypes with different growth characteristics. In environment, drought often develops during a growing season and occurs for a short period, which tolerant plants can manage to survive and complete their growth cycle. Drought resistance mechanisms can be understood by methods which simulate field-like conditions and quantify drought responses. Soil water deficit causing drought stress in crop plants has been tested in Arabidopsis using controlled soil moisture treatment. Controlled drought treatment, exposing plants to constant levels of soil moisture deficit, enables the evaluation between genotypes/ecotypes for plant responses to sublethal drought. Phenopsis is an alternative method for an automated controlled drought screen, which is used to compare the performance of different Arabidopsis ecotypes (accessions) and resulted in the identification of a resistant accession, An1 [99]. Controlled drought was also used to study the response of the Arabidopsis erecta mutant and ERECTA gene complementation [100], the overexpression of the Arabidopsis ESKIMO1 gene [101] and overexpression of the Pro biosynthesis gene in chickpea [102]. Comprehensive physiological and molecular studies have not yet been done on the response of plants to moderate drought (mDr). A transcriptome study in loblolly pine (Pinus taeda), treated for

http://dx.doi.org/10.5772/65246

cycles of mild drought and recovery [103], revealed a photosynthetic acclimation pattern in response to mild drought in contrast to photosynthesis inhibition under severe drought. A comprehensive understanding of the response of plants to mDr with physiological and molecular tools provided a better understanding of the acclimation process. A semi-automated, controlled mDr testing system was employed to compare with pDr treatment for physiological and molecular responses. This revealed differential gene reprogramming under the two drought treatments. The dissection of mDr treatment is presented using a time-course study to provide a picture of physiological and molecular responses toward acclimation in plant growth.

In recent years, much molecular information has been generated on the response of plants to environmental stresses. Plants respond to environmental stresses such as drought by the induction of both regulatory and functional sets of genes. Very little is known about the early events in the perception of stress signals. The common stress signaling pathways have been distinguished into abscisic acid (ABA) dependent and ABA independent. Most of the key genes in these pathways have been identified, such as transcription factors belonging to the class of DRE-binding protein (DREB)/C-repeat-binding factor (CBF), ABA-binding factor (ABF), MYC and MYB, including the identification of the stress-responsive cis-elements ABA-responsive element (ABRE) and dehydration responsive element. Downstream of the early signal perception events, signaling genes and molecules acting as secondary messengers have been identified, revealing the role of Ca^{+} and reactive oxygen species (ROS) as secondary messengers. These regulatory mechanisms induce downstream functional genes, which are needed to establish new cellular homeostasis that leads to drought tolerance and/or resistance. Most of our knowledge of drought responses at the molecular level is based on plant responses to molecular laboratory experimental conditions of dehydration and/or osmotic treatments. Laboratory conditions are far from the soil water deficit met by plants under field conditions, but these studies has provided valuable knowledge. Signaling pathways of ABA dependent and ABA independent have become a paradigm in plant biotic/abiotic stress responses [104]. These pathways were discovered in Arabidopsis (Arabidopsis thaliana) as a model system, which paved the way for the discovery of parallel pathways in other crop plants such as in rice (Oryza sativa) as a model for monocot plants. A number of drought treatments have been used to test the response of plants for improved tolerance/resistance. One method is progressive drought (pDr), in which water is with held for a certain period of time until symptoms of wilting are observed. Usually, this method of drought treatment has been used to determine survival rate or to monitor gene expression changes of wild-type plants or of plant genotypes overexpressing candidate genes for drought tolerance. These studies have helped to study plant responses to drought at both the physiological and molecular levels. However, one of the drawbacks of pDr treatment is that it cannot be used to compare the performance of different genotypes with different growth characteristics. In nature, drought often develops during a growing season and occurs for a short period, which tolerant plants can manage to survive and complete their growth cycle. Soil water deficit causing drought stress in crop plants has been tested in Arabidopsis using controlled soil moisture treatment. Controlled drought treatment, exposing plants to constant levels of soil moisture deficit, enables the evaluation between genotypes/ecotypes for plant responses to drought.

A large variety of stress responses in plants are influenced by ethylene metabolism and signaling. Ethylene signaling pathway is a major cross-link between ethylene and other plant hormones metabolism (ABA and GA). Interactions between ethylene and other plant hormones also benefit immediate stress responses such as stomatal closure as well as long term adaptations under severe drought conditions. Candidate genes that are related to maintenance of growth under low water conditions are agronomically important since they provide an efficient resource for crop improvement. A transgenic potato (*Solanum tuberosum*) cultivar, containing a betaine aldehyde dehydrogenase (*BADH*) gene from spinach (*Spinacia oleracea*) under the control of a stress induced *Arabidopsis* promoter, has been reported to exhibit improved growth after induction of BADH by NaCl and drought stress.

Proline an osmolyte in plants accumulates under different stress conditions. The amino acid proline in plant cells contribute to osmotic adjustments under adverse conditions. The enzyme Δ1-pyrroline-5-carboxylate synthetase (P5CS1) is a major component in proline biosynthesis. A study with P5CS1-deficient *Arabidopsis* mutants indicated that proline synthesis is required in order to maintain growth at low water availability [105]. Proline dehydrogenase 1 (PDH1)-deficient *Arabidopsis* mutants with blocked proline catabolism exhibited decreased root growth, fresh weight and dry weight. Additional components of the proline biosynthetic pathway are associated with stress responses. In transgenic soybean (*Glycine max*) with a Δ1-pyrroline-5-carboxylate reductase (*P5CR*) gene and the antisense construct from *Arabidopsis*, it was found that proline might enhance survival during drought stress [106]. The *P5CR* gene and antisense construct were manipulated using an inducible heat shock promoter (IHSP). Two transgenic potato lines, which expressed a trehalose-6-phosphate synthase (*TPS1*) gene from yeast (*Saccharomyces cerevisiae*) were found to be more effective in keeping water and acceptable levels of photosynthesis during drought compared to WT-plants [107]. The expression of aldehyde dehydrogenases (ALDHs) is upregulated under stress situations such as dehydration, salinity and oxidative stress. ALDHs are able to convert highly reactive aldehydes and hence extenuate oxidative stress [108]. Two drought tolerant Andean native potato clones (*Solanum tuberosum* subsp. *andigena*) under transcriptome analysis showed that aldehyde dehydrogenase family 7 (ALDH7) was induced under drought stress conditions [109]. Functional analyses of an ALDH7 gene member (*GmPP55*) from soybean (*Glycine max*) confirmed these studies. Transgenic *Arabidopsis* and tobacco (*Nicotiana tabacum*) plants exhibited improved tolerance to H_2O_2 as well as salt and drought conditions during different developmental stages [110].

5. MAS for drought tolerance

Conventional breeding for developing drought-tolerant crop varieties is time-consuming and labor intensive due to the quantitative nature of drought tolerance and difficulties in selection for drought tolerance. The identification of genomic regions associated with drought tolerance would enable breeders to develop improved cultivars with increased drought tolerance using marker-assisted selection (MAS). Plant breeding has benefited from DNA marker technologies that were used to establish saturated genetic maps in major crop species including cereals and legumes. Markers in a high density genetic map will allow the precise tagging of mono- and oligogenic traits, with the dual goal of marker-assisted selection for traits and positional cloning of the underlying genes. Use of genomic tools like molecular markers and other tools in integrated approach for crop improvement has also been referred as "genomics- assisted breeding". Mapping of genomes has been of interest to identify genomic locations of disease resistance genes and other yield-related traits. However, due to very low polymorphisms in few cultivated crops gene pool, progress in genomic research has been relatively slow compared with other highly polymorphic species. Important considerations for undertaking molecular breeding are molecular markers, genetic maps and markers associated with traits. During the early days of genomic studies, isozyme markers were used for map development in chickpea. Expression of these markers was influenced by the environment and their number was small. Restriction fragment length polymorphism (RFLP) and Random amplified polymorphic DNA (RAPD) markers were also used for genetic mapping studies. After development of simple sequence repeat (SSR) or microsatellite markers, the use of molecular markers increased extensively. SSR markers were considered as the marker of choice in plant breeding due to their multi-allelic and co-dominant nature. Several hundred SSR markers have been developed from genomic DNA libraries. In several cases, the mapping populations used for developing the maps were also phenotyped for the segregating traits. Analysis of phenotyping data together with genotyping data in some cases identified molecular markers associated with the genes/quantitative trait loci (QTLs), controlling resistance to key diseases (ascochyta blight, fusarium wilt, botrytis grey mold, rust), morphological traits (single pod *vs.* double pod, flowering time and flower color), seed yield and yield components, *etc.* Marker-assisted breeding reduces the effect of environmental conditions during the selection process, which is a major hindrance in conventional breeding under drought.

5.1. QTL analysis for drought tolerance in chickpea

Compared to the conventional breeding approaches for improved productivity under water limited environments, genomics offers great opportunities for dissecting quantitative traits into their single genetic determinants. The release of varieties through conventional breeding approaches is coupled with identification of several large-effect QTLs for grain yield under drought in different crops. Independent and epistatic QTLs for grain yield and other traits of agronomic importance were studied in different crops. Only a few studies reported major QTLs affecting yield advantage under both drought stress and non-stress environments. Drought is normally associated with increased incidence of diseases such as blast, brown spot, and bacterial blight. Few studies have been undertaken to understand the genetics of these abiotic and biotic stresses simultaneously in a mapping population. Identification of QTLs is paving the way to MAS and assisted pyramiding of the beneficial QTL alleles. Markers can be used in marker-assisted selection (MAS) for improving the desired trait. Isolation and validation of genes underlying the QTL/genes for the traits of interest is an essential step to determine gene function. QTLs for drought tolerance have been identified for several major and important crop species such as rice, maize, wheat, barley, sorghum, pearl millet, soybean and chickpea. These QTLs were identified for a variety of important traits including: (1) yield and yield-contributing traits under water-deficit conditions (in the case of wheat, maize, rice, soybean and pearl millet), (2) physiological responses including water-soluble carbohydrates, carbon isotope ratio, osmotic potential, chlorophyll content, flag leaf rolling index, grain carbon isotope discrimination, relative water content, leaf osmotic potential, osmotic adjustment, chlorophyll and chlorophyll fluorescence parameters to drought stress (in the case of wheat, maize and rice), (3) flowering time including anthesis to silking interval (in maize), (4) root traits (rice, maize, wheat, soybean and chickpea), (5) stay green (sorghum) and (6) nitrogen fixation (soybean). When the QTLs identified for drought tolerance traits contribute higher phenotypic variation, they are considered major QTLs. These QTLs, after validation in desired germplasm, can be used for introgressing drought tolerance from the donor genotypes (generally used for identification of the QTL for the trait) into elite, less drought-tolerant cultivars or breeding lines (recipient parents) without transfer of undesirable or deleterious genes from the donors (linkage drag). After identifying important QTLs, the next step involves the identification of candidate sequences, validate their role and proceed with the direct manipulation using the gene itself as marker for MAS. In chickpea the RILs of ICC 4958 × Annigeri have been extensively studied for root traits. An SSR marker (TAA 170) was identified for a major QTL that accounted for 33% of the variation for root weight and 33% of the variation for root length [111]. Recent preliminary screening of the chickpea mini-core germplasm collection for root proliferation and depth in cylinder culture indicated that contrasting parents are available with wider variation for these traits than that present between ICC 4958 and Annigeri [112]. Nayak et al. [113] undertook identification of QTLs and genes for drought tolerance using linkage mapping and association mapping approaches in Chickpea (*Cicer arietinum*).SSR markers were tested for polymorphism on parental genotypes of the inter-specific (ICC 4958 × PI 489777) and intra-specific mapping population (ICC 4958 × ICC 1882). As a result, a comprehensive inter-specific genetic map of 621 marker loci, spanning a genetic distance of 984.11cM was prepared. Varshney et al. [114] identified genomics and physiological approaches for root trait breeding to improve drought tolerance in Chickpea (*Cicer arietinum* L.). Molecular markers and candidate genes associated with root traits are being targeted to introgress the QTLs for root traits from drought-tolerant genotypes to drought-sensitive genotypes following marker-assisted breeding strategies. Varshney et al. (2014) reported a "*QTL-hotspot*" (ICCM0249, NCPGR127, TAA170, NCPGR21, TR11, GA24 and STMS11) on CaLG04 in the chickpea genome, identified in analysis on both RIL populations, (ICCRIL03 (ICC 4958 × ICC1882) and ICCRIL04 (ICC 283 × ICC 8261) that contain 45 M-QTLs and 973 E-QTLs for several drought tolerance traits contributing up to 58.20% phenotypic variation for targeted traits [22, 115].

In the last 20 years, considerable progress has been made towards mapping QTLs for drought resistance traits in rice however, there have been few successful cases of their application in MAB. The success rate of using QTLs in molecular breeding reflects the lack of repeatability of QTL effects across genetic backgrounds and environments. In recent years, several researchers developed mapping populations between high-yielding lines (IR64, Swarna and MTU1010) and drought-tolerant local landraces and wild cultivars to map grain yield QTLs for reproductive stage-specific drought stress.

To the best our knowledge, none of the studies were conducted under natural drought conditions predominant in tough environments (TEs) and these QTLs were identified in moderate stress environment (MSE) and QTLs mapped under severe drought stress conditions. Successful marker-assisted selection to improve yield mainly relied on the use of high yielding lines to identify large-effect QTLs and evaluation of their consistent effects. Studies in MSE may limit the chances of detecting QTLs for drought resistance that are widely applicable to target populations of environments, as the timing and intensity of stress vary over years in rain fed rice ecosystems, which ultimately changes the plants' responses and traits involved in drought-resistance mechanisms. Most of the indica × indica derived rice lines used in QTL mapping of drought resistance were not adapted to TEs. The importance of field experiments in TPEs to identify QTLs for rice yield under natural drought stress was emphasized. Recombinant inbred lines (RILs) derived from locally adapted indica rice lines to detect QTLs for plant production traits under drought stress in TPEs, but no yield QTL was identified.

Quantitative genetics, with wide range of molecular markers available, provide identification of the genetic factors (quantitative trait loci-QTLs) responsible for expression of traits. Recent development in molecular marker technology is expected to enable greater power in detection of QTL for agronomically important traits and utilization of QTL information for crop improvement. Thus marker-assisted selection could significantly enhance in improving crop drought tolerance, if QTL with significant effects can be identified.

6. Conclusion

Environment change is a universal phenomenon that has started to have adverse impact on agriculture. The global temperature is predicted to rise by 2.5–4.3°C by the end of the century. The situation is further likely to get worse due to the occurrence of increase in the irregularity of rainfall, drought, flood and land degradation. With predicted climate change scenarios and continuous population explosion, there is a great need to develop high-yielding varieties with improved drought tolerance. Breeding for drought tolerance is not simple. Under a particular environment, some physiological or metabolic processes can be modified through breeding, either as single traits or as a combination of traits. Optimal drought-adaptation requires the combination of several morphological, physiological and phenological processes which depends on multiple genes and varies within each target environment. Conventional and marker-based approaches coupled with each other have been used for drought tolerance. The conventional breeding approach is based on selection for yield and its components in a given drought environment. But this approach requires large investments in land, labor and capital to screen a large number of progenies plus the difficulty of sampling even a part of the expected range of variability in stress occurrence in the target environment hence this approach is not successful. Traits including modification of the root system, stomatal control, and leaf area, as well as matching plant phenology with the environment, could help in improving productivity under drought stress conditions. Recent research breakthroughs in biotechnology have revived interest in targeted drought tolerance breeding and use of new genomics tools to increase crop productivity.

Marker-assisted breeding is an important technique for improvement of crop productivity against drought stress. As a complement to the recent rapid progress in genomics, a better understanding of physiological mechanisms of drought response contributes to the progress of crop productivity against drought tolerance. Mostly physiological traits associated with drought tolerance are quantitative in nature. An important research strategy that has been widely used over the past two decades to deal with such complexity is the use of molecular markers to identify quantitative trait loci (QTLs) in appropriate mapping populations. Once molecular markers (i.e. for trait QTLs) linked to specific drought tolerance component traits found, it is possible to move them into adapted cultivars or other agronomic backgrounds through marker-assisted breeding. Moreover, in adapted genotypes identification of QTLs for the key traits responsible for improved productivity under drought could be helpful in accelerating the process of pyramiding of favorable alleles for better yield and production. Integration of knowledge from plant physiology and biotechnology into plant breeding can help developing best cultivars for drought tolerance. The availability of a large number of molecular markers, dense genetic maps, and markers associated with traits and transcriptomics resources have made it possible to integrate genomics technologies into chickpea improvement. Understanding plant response to water stress for key drought stress traits and screening of mapping populations for these traits for QTL identification are of prime importance for future drought stress breeding. Food security requires investments in this domain, in particular with new genotypes that can at least maintain an acceptable productivity under drought stress condition.

Author details

Summy Yadav[1] and Kamal Dutt Sharma[2*]

*Address all correspondence to: kdutt1966@gmail.com

1 Ahmedabad University, Gujarat, India

2 CCS Haryana Agriculture University, Haryana, India

References

[1] Gaspar, T., Franck, T., Bisbis, B., Kevers, C., Jouve, L., Hausman, J.F., Dommes, J. 2002. Concepts in plant stress physiology. Application to plant tissue cultures. Plant Growth Regul., 37: 263–285.

[2] Jaleel, C.A., Manivannan, P., Sankar, B., Kishorekumar, A., Gopi, R., Somasundaram, R., Panneerselvam, R. 2007. Water deficit stress mitigation by calcium chloride in

Catharanthus roseus; effects on oxidative stress, proline metabolism and indole alkaloid accumulation. Colloids Surf. B: Biointerfaces, 60: 110–116.

[3] Zhu, J.K. 2002. Salt and drought stress signal transduction in plants. Annu. Rev. Plant Biol., 53: 247–273.

[4] Razmjoo, K., Heydarizadeh, P., Sabzalian, M.R. 2008. Effect of salinity and drought stresses on growth parameters and essential oil content of Matricaria chamomile. Int. J. Agric. Biol., 10: 451–454.

[5] Blum, A. 1996. Constitutive traitsaffecting plant performance under stress. In: Edmeades, G.O., Banziger, M., Mickelson, H.R., Pena-Valdivia, C.B. (eds.), Developing Drought and Low N Tolerant Maize, Proceedings of the Symposium. CIMMYT: Mexico D.F. pp. 131–135.

[6] Jaleel, C.A., Manivannan, P., Sankar, B., Kishorekumar, A., Gopi, R., Somasundaram, R., Panneerselvam, R. 2007. Induction of drought stress tolerance by ketoconazole in Catharanthus roseus is mediated by enhanced antioxidant potentials and secondary metabolite accumulation. Colloids Surf. B: Biointerfaces, 60: 201–206.

[7] Smirnoff, N. 1993. The role of active oxygen in the response of plants to water deficit and desiccation. New Phytol., 125: 27–58.

[8] Jaleel, C.A., Gopi, R., Sankar, B., Gomathinayagam, M., Panneerselvam, R. 2008. Differential responses in water use efficiency in two varieties of Catharanthus roseus under drought stress. Comp. Rend. Biol., 331: 42–47.

[9] Jaleel, C.A., Manivannan, P., Lakshmanan, G.M.A., Gomathinayagam, M., Panneerselvam, R. 2008. Alterations in morphological parameters and photosynthetic pigment responses of Catharanthus roseus under soil water deficits. Colloids Surf. B: Biointerfaces, 61: 298–303.

[10] Martinez, J.P., Silva, H., Ledent, J.F., Pinto, M. 2007. Effect of drought stress on the osmotic adjustment, cell wall elasticity and cell volume of six cultivars of common beans (Phaseolus vulgaris L.). Eur. J. Agron., 26: 30–38.

[11] Chaves, M.M., Pereira, J.S., Maroco, J., Rodriques, M.L., Ricardo, C.P.P., Osorio, M.L., Carvatho, I., Faria, T., Pinheiro, C. 2002. How plants cope with water stress in the field photosynthesis and growth? Ann. Bot., 89: 907–916.

[12] Reddy, A.R., Chaitanya, K.V., Vivekanandan, M. 2004. Drought induced responses of photosynthesis and antioxidant metabolism in higher plants. J. Plant Physiol., 161: 1189–1202.

[13] Price, A.H., Cairns, J.E., Horton, P., Jones, H.G., Griffiths, H. 2002. Linking drought-resistance mechanisms to drought avoidance in upland rice using a QTL approach: progress and new opportunities to integrate stomatal and mesophyll responses. J. Exp. Bot., 53: 989–1004.

[14] Ingram, J., Bartels, D. 1996. The molecular basis of dehydration tolerance in plants. Annu. Rev. Plant. Physiol. Plant Mol. Biol., 47: 377–403.

[15] Yancey, P.H., Clark, M.E., Hand, S.C., Bowlus, R.D., Somero, G.N. 1982. Living with water stress: evolution of osmolyte systems. Science, 217: 1214–1222.

[16] Ramanjulu, S., Bartels, D. 2002. Drought- and desiccation-induced modulation of gene expression in plants. Plant Cell. Environ., 25: 141–151.

[17] Bartels, D., Sunkar, R. 2005. Drought and salt tolerance in plants. Crit. Rev. Plant Sci., 24: 23–58.

[18] Urao, T., Yakubov, B., Satoh, R., Yamaguchi-Shinozaki, K., Seki, M., Hirayama, T., Shinozaki, K. 1999. A transmembrane hybrid-type histidine kinase in Arabidopsis functions as an osmosensor. Plant Cell., 11: 1743–1754.

[19] Ueguchi, C., Koizumi, H., Suzuki, T., Mizuno, T. 2001. Novel family of sensor histidine kinase genes in Arabidopsis thaliana. Plant Cell. Physiol., 42: 231–235.

[20] Shinozaki, K., Yamaguchi-Shinozaki, K. 1997. Gene expression and signal transduction in water-stress response. Plant Physiol., 115: 327–334.

[21] Shinozaki, K., Yamaguchi-Shinozaki, K. 2007. Gene networks involved in drought stress response and tolerance. J. Exp. Bot., 58: 221–227.

[22] Varshney, R.K., Hoisington, D.A., Tyagi, A.K. 2006. Advances in cereal genomics and applications in crop breeding. Trends Biotechnol., 24: 490–499.

[23] Varshney, R.K., Graner, A., Sorrells, M.E. 2005. Genic microsatellite markers in plants: features and applications. Trends Biotechnol., 23: 48–55.

[24] Varshney, R.K., Close, T.J., Singh, N.K., Hoisington, D.A., Cook, D.R. 2009. Orphan legume crops enter the genomics era! Curr. Opin. Plant Biol., 12: 202–210.

[25] Glaszmann, J.C., Kilian, B., Upadhyaya, H.D., Varshney, R.K. 2010. Accessing genetic diversity for crop improvement. Curr. Opin. Plant Biol., 13: 1–7.

[26] Ribaut, J.M., Jiang, C., Gonzalez-de-Leon, D., Edmeades, G.O., Hoisington, D.A. 1997. Identification of quantitative trait loci under drought conditions in tropical maize. 2. Yield components and marker assisted selection strategies. Theor. Appl. Genet., 94: 887–896.

[27] Rehman, A.U., Malhotra, R.S., Bett, K., Tar'an, B., Bueckert, R., Warkentin, T.D. 2011. Mapping QTL associated with traits affecting grain yield in chickpea (Cicer arietinum L.) under terminal drought stress. Crop Sci. 2010.03.0129: 51:450–463.

[28] Croser, J.S., Ahmad, F., Clarke, H.J., Siddique, K.H.M. 2003. Utilization of wild Cicer in chickpea improvement- Progress, constraints, and prospects. Aust. J. Agr. Res., 54: 429–444.

[29] Singh, N.H., Dhaliwal, G.S. 1972. Effect of soil temperature on seedling emergence in different crops. Plant Soil., 37: 441–444.

[30] Ellis, R.H., Covell, S., Roberts, E.H., Summerfield, R.J. 1986 The influence of temperature on seed germination rate in grain legumes. J. Exp. Bot., 37: 1503–1515.

[31] Wery, J., Turc, O., Lecoeur, J. 1993. Mechanism of resistance to cold, heat and drought in cool-season legumes, with special reference to chickpea and pea. In: Singh, K.B., Saxena, M.C. (eds.) Food Legumes. Wiley Publishing: Chichester, UK. pp. 271–291.

[32] Turner, N.C. 1979. Drought resistance and adaptation to water deficit in crop plants. In Mussell, H., Staples, R.C. (eds.), Stress Physiology in Crop Plants. pp: 343–367. New York.

[33] Toker, C., Canci, H. 2006. Selection for drought and heat resistance in chickpea under terminal drought conditions. In Kharkwal, M.C. (ed.) Food Legumes for Nutritional Security and Sustainable Agriculture. 4th International Food Legumes Research Conference. pp. 18–22. Indian Agricultural Research Institute: New Delhi.

[34] Toker, C., Llunch, C., Tejera, N.A., Serraj, R., Siddique, K.H.M. 2007. Abiotic stresses. In Yadav, S.S., Redden, R.J., Chen, W., Sharma, B. (eds.) Chickpea Breeding and Management. pp. 474–496. CAB International: Wallingford, UK.

[35] Araus, J.L., Slafer, G.A., Reynolds, M.P., Royo, C. 2002. Plant breeding and drought in C-3 cereals: what should we breed for? Ann. Bot., 89: 925–940.

[36] Siddique, K.H.M., Loss, S.P., Thomson, B.D. 2003. Cool season grain legumes in dryland Mediterranean environments of Western Australia: significance of early flowering. In: Saxena, N.P. (ed.) Management of Agricultural Drought. Science Publishers: Enfield, NH. pp. 151–161.

[37] Kumar, J., Abbo, S. 2001. Genetics of flowering time in chickpea and its bearing on productivity in the semi-arid environments. Adv. Agron., 72: 107–138.

[38] DaMatta, F.M. 2004. Exploring drought tolerance in coffee: a physiological approach with some insights for plant breeding, Braz. J. Plant Physiol., 16: 1–6.

[39] Turner, N.C., Wright, G.C., Siddique, K.H.M. 2001 Adaptation of grain legumes (pulses) to water-limited environments. Adv. Agron., 71: 193–231.

[40] Tuberosa, R., Sanguinet, M.C., Landi, P., Guiliani, M.M., Salvi, S., Conti, S. 2002. Identification of QTLs for root characteristics of maize hydroponics analysis of their overlap with QTLs for grain yield in field at water regimes. Plant Mol. Biol., 48: 697–712.

[41] Saxena, N.P. 2003 Management of drought in chickpea-holistic approach. In: Saxena, N.P. (eds.) Management of Agricultural Drought-Agronomic and Genetic Options. Oxford and IBH Publishing Co. Pvt. Ltd.: New Delhi, India. pp. 103–122.

[42] Krishnamurthy, L., Kashiwagi, J., Upadhyaya, H.D., Serraj, R. 2003 Genetic diversity of drought-avoidance root traits in the mini-core germplasm collection of chickpea. Int. Chick. Pigeonpea Newslett., 10: 21–24.

[43] Davies, W., Zhang, J. 1991. Root signals and the regulation of growth and development of plants in dry soil. Ann. Rev. Plant Physiol. Plant Mol. Biol., 42: 55–76.

[44] Richards, R.A., Rawson, H.M., Johnson, D.A. 1986. Glaucousness in wheat: its development, and effect on water-use efficiency, gas exchange and photosynthetic tissue temperatures. Aust. J. Plant Physiol., 13: 465–473.

[45] Ludlow, M.M., Muchow, R.C. 1990. A critical evaluation of traits for improving crop yields in water-limited environments. Adv. Agron., 43: 107–153.

[46] Shashidar, R., Saroj, S. D., Harjare, S., Dhokane, V., Sharma, A., Bandekar, J. R. 2000. Tropics. Inheritance of flower colour in chickpea. J. Hered., 91: 5.

[47] Ugherughe, P.O. 1986. Relationship between digestibility of Bromus inermis plant parts. J. Agron. Crop Sci., 157: 136–143.

[48] Close, T.J. 1997. Dehydrins: a commonality in the response of plants to dehydration and low temperature. Physiol. Plant., 100: 291–296.

[49] Hara, M., Terashima, S., Kuboi. T. 2001. Characterization and cryoprotective activity of cold-responsive dehydrin from Citrus unshiu. J. Plant Physiol., 158: 1333–1339.

[50] Hincha, D.K., Hagemann, M. 2004. Stabilization of model membranes during drying by compatible solutes involved in the stress tolerance of plants and microorganisms. Biochem. J., 383: 277–283.

[51] Hung, S.H., Yu, C.W. 2005. Hydrogen peroxide functions as a stress signal in plants. Bot. Bull. Acad. Sinica., 41: 1–10.

[52] Chen, Z., Gallie, D.R. 2004. The ascorbic acid redox state controls guard cell signaling and stomatal movement. Plant Cell., 16: 1143–1162.

[53] Saxena, M.C. 2003. Recent advances in chickpea agronomy. In: Proceeding of the International Workshop on Chickpea Improvement, 28th Feb.-3rd Mar. 1979, ICRISAT Centre, India. International Crops Research Institute for the Semi – Arid Tropics: Patancheru, India. pp. 29–31.

[54] Gong, H., Zhu, X., Chen, K., Wang, S., Zhang, C. 2005. Silicon alleviates oxidative damage of wheat plants in pots under drought. Plant Sci., 169: 313–321.

[55] Hasegawa, P.M., Bressan, R.A., Zhu, J.K., Bohnert, H.J. 2000. Plant cellular and molecular responses to high salinity. Annu. Rev. Plant Phys., 51: 463–499.

[56] Farooq, M., Basra, S.M.A., Ahmad, N. 2007. Improving the performance of transplanted rice by seed priming. Plant Growth Regul., 51: 129–137.

[57] Fazeli, F., Ghorbanli, M., Niknam, V. 2007. Effect of drought on biomass, protein content, lipid peroxidation and antioxidant enzymes in two sesame cultivars. Biol. Plant., 51: 98–103.

[58] Pastori, G., Foyer, C.H., Mullineaux, P. 2000. Low temperature-induced changes in the distribution of H_2O_2 and antioxidants between the bundle sheath and mesophyll cells of maize leaves. J. Exp. Bot., 51: 107–113.

[59] Chew, O., Whelan, J., Miller, A.H. 2003. Molecular definition of the ascorbate-glutathione cycle in Arabidopsis mitochondria reveals dual targeting of antioxidant defences in plants. J. Biol. Chem., 278: 46869–46877.

[60] Lima, A.L.S., DaMatta, F.M., Pinheiro, H.A., Totola, M.R., Loureiro, M.E. 2002. Photochemical responses and oxidative stress in two clones of Coffea canephora under water deficit conditions. Environ. Exp. Bot., 47: 239–247.

[61] Pinheiro, H.A., DaMatta, F.M., Chaves, A.R.M., Fontes, E.P.B., Loureiro, M.E. 2004. Drought tolerance in relation to protection against oxidative stress in clones of Coffea canephora subjected to long-term drought. Plant Sci., 167: 1307–1314.

[62] Hasegawa, P.M., Bressan, R.A., Zhu, J.K., Bohnert, H.J. 2000. Plant cellular and molecular responses to high salinity. Annu. Rev. Plant Phys., 51: 463–499.

[63] Havaux, M. 1998. Carotenoids as membrane stabilizers in chloroplasts. Trends in Plant Sci., 3: 147–151.

[64] Wahid, A., Gelani, S., Ashraf, M., Foolad, M.R. 2007. Heat tolerance in plants: an overview. Environ. Exp. Bot., 61: 199–223.

[65] Morgan, P.W. 1990. Effects of abiotic stresses on plant hormone systems, In: Stress Responses in Plants: Adaptation and Acclimation Mechanisms. Wiley-Liss, NewYork. Inc. pp. 113–146.

[66] Taiz, L., Zeiger, E. 2006. Plant Physiology, 4th Ed., Sinauer Associates Inc. Publishers: MA.

[67] Nilsen, E.T., Orcutte, D.M. 1996. Phytohormones and plant responses to stress. In: Nilsen, E.T., Orcutte, D.M. (eds.) Physiology of Plant under Stress: Abiotic Factors. John Wiley and Sons: New York. pp. 183–198.

[68] Ludwig-Müller, J. 2007. Indole-3-butyric acid synthesis in ecotypes and mutants of Arabidopsis thaliana under different growth conditions. J. Plant Physiol., 164: 47–59.

[69] Sadiqov, S.T., Akbulut, M., Ehmedov, V. 2002. Role of Ca^{2+} in drought stress signaling in wheat seedlings. Biochemistry-Moscow, 67: 491–497.

[70] Vartanian, N., Marcotte, L., Ciraudat, J. 1994. Drought Rhizogenesis in Arabidopsis thaliana: differential responses of hormonal mutants. Plant Physiol., 104: 761–767.

[71] Taylor, I.B. 1991. Genetics of ABA synthesis. In: Davies, W.J., Jones, H.G., (eds.) Abscisic Acid: Physiology and Biochemistry. Bios Scientific Publishers Ltd.: UK. pp. 23–38.

[72] Turner, N.C., Wright, G.C., Siddique, K.H.M. 2001. Adaptation of grain legumes (pulses) to water-limited environments. Adv. Agron., 71: 123–231.

[73] Ball, R.A., Oosterhuis, D.M., Mauromoustakos, A. 1994. Growth dynamics of the cotton plant during water-deficit stress. Agron. J., 86: 788–795.

[74] Young, T.E., Meeley, R.B., Gallie, D.R. 2004. ACC synthase expression regulates leaf performance and drought tolerance in maize. Plant J., 40: 813–825.

[75] Pierik, R., Sasidharan, R., Voesenek, L.A.C.J. 2007. Growth control by ethylene: adjusting phenotypes to the environment. J. Plant Growth Regul., 26: 188–200.

[76] Kasukabe, Y., He, L., Nada, K., Misawa, S., Ihara, I., Tachibana, S. 2004. Overexpression of spermidine synthase enhances tolerance to multiple environmental stresses and up-regulates the expression of various stress-regulated genes in transgenic Arabidopsis thaliana. Plant Cell. Physiol., 45: 712–722.

[77] Bouchereau, A., Aziz, A., Larher, F., Tanguy, M. 1999. Polyamines and environmental challenges. Rec. Develop, Plant Sci., 140: 103–125.

[78] Wan, B., Lin, Y., Mou, T. 2007. Expression of rice Ca(2+)-dependent proteinkinases (CDPKs) genes under different environmental stresses. FEBS Lett., 581: 1179–1189.

[79] Bohnert, H.J., Nelson, D.E., Jensen, R.G. 1995. Adaptations to environmental stresses. Plant Cell., 7: 1099–1111.

[80] Bray, E. 1997. Plant responses to water deficit. Trends Plant Sci., 2: 48–54.

[81] Ackerly, D.D., Dudley, S.A., Sultan, S.E. 2000. The evolution of plant ecophysiological traits: recent advances and future directions. Bioscience, 50: 979–995.

[82] Schulze, E.D., Robichaux, R.H., Grace, J., Rundel, P.W., Ehleringer, J.R. 1987. Plant water balance. Bioscience, 37: 30–37.

[83] Kashiwagi, J., Krishnamurthy, L., Crouch, J.H., Serraj, R. 2006. Variability of root length density and its contributions to seed yield in chickpea (Cicer arietinum L.) under terminal drought stress. Field Crops Res., 95: 171–181.

[84] Sabaghpour, S.H., Kumar, J. 2002. Role of initial growth vigor in drought escape. In: Proceeding of Seventh International Conference on Development and Management of Dry Lands in the 21st Century, Iran. pp. 57–58.

[85] Sabaghpour, S.H., Kumar, J., Rao, T.N. 2003. Inheritance of growth vigor and its association with other characters in chickpea. Plant Breed., 122: 542–544.

[86] Upadhyaya, H.D., Kashiwagi, J., Varshney, R.K., Gaur, P.M., Saxena, K.B., Krishnamuthy, L., Gowda, C.L.L., Pundir, R.P.S., Basu, P.S., Singh, I.P. 2012. Phenotyping chickpeas and pigeonpea for adaptation to drought. Front. Physiol., 3: 179

[87] Ulemale, C.S., Mate, S.N., Deshmukh, D.V. 2013. Physiological indices for drought tolerance in chickpea (Cicer arietinum L.). World J. Agric. Sci., 9(2): 123–131.

[88] Levitt, J. 1980. Response of Plants to Environmental Stress. Vol. 2. Academic Press: New York.

[89] Gupta, S.C., Rathore, A.K., Sharma, S.N., Saini, R.S. 2000. Response of chickpea cultivars to water stress. Indian J. Plant Physiol., 5: 274–276.

[90] Sharma, S.N., Singh, G. 1989. Limitations of gas exchanges in intact leaves during ontogeny of field grown chickpea. J. Exp. Bot., 40: 1399–1406.

[91] Kushwaha, S.R., Deshmukh, P.S., Singh, T. 2003. Studies on drought tolerance in chickpea- physiological traits yield and yield component. Indian J. Plant Physiol., (Special Issue): 386–389.

[92] Omae, H., Kumar, A., Egawa, Y., Kashiwaba, K., Shono, M. 2005. Midday drop of leaf water content related to drought tolerance in snap bean (Phaseolus vulgaris L.). Plant Prod Sci., 8(4): 465–467.

[93] Nayyar, H., Gupta, D. 2006. Differential sensitivity of C3 and C4 plants to water deficit stress: association with oxidative stress and antioxidants. Environ. Exp. Bot., 58: 106–113.

[94] Blum, A., Ebercon, A. 1981. Cell membrane stability as a measure of drought and heat tolerance in wheat. Crop Sci., 21: 43–47.

[95] Xu, R.Q., Sun, Q.X., Zhang, S.Z. 1997. Screening methods and indices of heat tolerance in spring wheat. Acta Agril. Boreali Sinic. 12(3):22-29.

[96] Pouresmael, M., Nejad, R.A.K., Mozafari, J., Najafi, F., Moradi, F. 2013. Efficiency of screening criteria for drought tolerance in chickpea. Archv. Agron. Soil Sci., 59: 1675–1693.

[97] Kumar, N., Nandwal, A.S., Waldia, R.S., Singh, S., Devi, S., Sharma, K.D., Kumar, A. 2012. Drought tolerance in chickpea as evaluated by root characteristics, plant water status, membrane integrity and chlorophyll fluorescence techniques. Expl. Agric., 48(3): 378–387.

[98] Farooq, M., Wahid, A., Kobayashi, N., Fujita, D., Barsa, S.M.A. 2009. Plant drought stress: effects, mechanisms and management. Agronomy for sustainable development. 29:185-212.

[99] Granier, C., Aguirrezabal, L., Chenu, K., Cookson, S.J., Dauzat, M., Hamard, P., Thioux, J.J., Rolland, G., Bouchier-Combaud, S., Lebaudy, A. 2006. PHENOPSIS, an automated platform for reproducible phenotyping of plant responses to soil water deficit in Arabidopsis thaliana permitted the identification of an accession with low sensitivity to soil water deficit. New Phytol., 169: 623–635.

[100] Masle, J., Gilmore, S.R., Farquhar, G.D. 2005. The ERECTA gene regulates plant transpiration efficiency in Arabidopsis. Nature, 436: 866–870.

[101] Bouchabke-Coussa, O., Quashie, M.L., Seoane-Redondo, J., Fortabat, M.N., Gery, C., Yu, A., Linderme, D., Trouverie, J., Granier, F., Te'oule' E. 2008. ESKIMO1 is a key gene involved in water economy as well as cold acclimation and salt tolerance. BMC Plant Biol., 8: 1–27.

[102] Bhatnagar-Mathur, P., Vadez, V., Devi, M., Lavanya, M., Vani, G., Sharma, K. 2009. Genetic engineering of chickpea (Cicer arietinum L.) with the P5CSF129A gene for osmoregulation with implications on drought tolerance. Mol. Breed., 23: 591–606.

[103] Va'squez-Robinet, C., Watkinson, J.I., Sioson, A.A., Ramakrishnan, N., Heath, L.S., Grene, R. 2010. Differential expression of heat shock protein genes in preconditioning for photosynthetic acclimation in water-stressed loblolly pine. Plant Physiol. Biochem., 48: 256–264.

[104] Shinozaki, K., Yamaguchi-Shinozaki, K. 2007. Gene networks involved in drought stress response and tolerance. J. Exp. Bot., 58: 221–227.

[105] Sharma, S., Villamor, J.G., Verslues, P.E. 2011. Essential role of tissue-specific proline synthesis and catabolism in growth and redox balance at low water potential. Plant Physiol., 157: 292–304.

[106] De Ronde, J.A., Spreeth, M.H., Cress, W.A. 2000. Effect of antisense L-Δ1-pyrroline-5-carboxylate reductase transgenic soybean plants subjected to osmotic and drought stress. Plant Growth Regul., 32: 13–26.

[107] Stiller, I., Dulai, S., Kondrák, M., Tarnai, R., Szabó, L., Toldi, O., Bánfalvi, Z. 2008. Effects of drought on water content and photosynthetic parameters in potato plants expressing the trehalose-6-phosphate synthase gene of Saccharomyces cerevisiae. Planta, 227: 299–308.

[108] Singh, S., Brocker, C., Koppaka, V., Chen, Y., Jackson, B.C., Matsumoto, A., Thompson, D.C., Vasiliou, V. 2013. Aldehyde dehydrogenases in cellular responses to oxidative/electrophilic stress. Free Radic. Biol. Med., 56: 89–101.

[109] Schafleitner, R., Rosales, R.O.G., Gaudin, A., Aliaga, C.A.A., Martinez, G.N., Marca, L.R.T., Bolivar, L.A., Delgado, F.M., Simon, R., Bonierbale, M. 2007. Capturing candidate drought tolerance traits in two native Andean potato clones by transcription profiling of field grown plants under water stress. Plant Physiol. Biochem., 45: 673–690.

[110] Rodrigues, S.M., Andrade, M.O., Gomes, A.P.S., Damatta, F.M., Baracat-Pereira, M.C., Fontes, E.P.B. 2006. Arabidopsis and tobacco plants ectopically expressing the soybean antiquitin-like ALDH7 gene display enhanced tolerance to drought, salinity, and oxidative stress. J. Exp. Bot. 57: 1909–1918.

[111] Chandra, S., Buhariwalla, H.K., Kashiwagi, J., Hari Krishna, S., Rupa Sridevi, K., Krishnamurthy, L., Serraj, R., Crouch, J.H. 2004. Identifying QTL-linked markers in

marker-deficient crops. In: Proceedings of 4th Int Crop Science Congress, Sept 26-Oct 1. 2004, Brisbane, Australia.

[112] Krishnamurthy, L. 2003. Genetic diversity of drought-avoidance root traits in the mini-core germplasm collection of chickpea. Int. Chick. Pigeonpea News, 10: 21.

[113] Nayak, S.N. 2010. Identification of QTLS and genes for drought tolerance using linkage mapping and association mapping approaches in chickpea (Cicer arietinum). PhD thesis, Osmania University, Hyderabad, India.

[114] Varshney, R.K., Bansal, K.C., Aggarwal, P.K., Datta, S.K., Craufurd, P.Q. 2011. Agricultural biotechnology for crop improvement in a variable climate: hope or hype? Trends Plant Sci., 16(7): 363–371.

[115] Gunes, A., Inal, A., Adak, M.S., Bagci, E.G., Cicek, N., Eraslan, F. 2008. Effect of drought stress implemented at pre- or post-anthesis stage some physiological as screening criteria in chickpea cultivars. Rus. J. Plant. Physiol., 55: 59–67.

Chapter 11

Barley Phenology: Physiological and Molecular Mechanisms for Heading Date and Modelling of Genotype-Environment-Management Interactions

Ahmed Ibrahim, Matthew Harrison,
Holger Meinke and Meixue Zhou

Additional information is available at the end of the chapter

http://dx.doi.org/10.5772/64827

Abstract

Barley heading date is important in adapting barley genotypes to different environments. Important factors affecting heading date in barley are temperatures, photoperiod and sowing date. Sowing date is a management option to influence heading date. It is used to reduce climatic risks such as frosts and water stress at sensitive periods during crop development. Three major genes control heading date in barley. These genes regulate photoperiod (Ppd-H1 and Ppd-H2), vernalization (Vrn- H1, Vrn-H2 and Vrn-H3) and the duration of the vegetative phase (Eps). The Ppd-H1 locus on chromosome 2(2H) regulates flowering time under long days. Ppd-H2 on 2H regulates phenology under short day length. Vernalization is mainly controlled by three loci (VRN-H1, VRN-H2 and VRN-H3), which interact in an epistatic fashion to determine vernalization sensitivity. Appropriate physiological and simulation frameworks such as that of APSIM-Barley are required to complement breeding efforts in order to determine the location of the Eps genes and describe and quantify the effects of environment and management on gene expression and their impact on yields and quality in barley.

Keywords: barley, photoperiod, vernalization, earliness per se, modelling

1. Introduction

World barley production is projected to reach 140 million metric tons (MMT) on 50 m hectares by 2016/2017 [1] with its greatest economic impact being related to its use as feed, food and for malting. The demand is projected to reach 142 MMT by 2050 [1]. It is therefore one of the major sources of income for the countries where it can be produced. Australia is the fourth largest barley grower in the world, producing about 8.7 MMT of barley in 2015 and contributes up to 30% of the world supply [2,3]. The supplies comprise 2.5 MMT of malting barley and 4.5 MMT of the feed barley [4, 5]. Currently, one-third of the world production is used for malting [6]. The grain is also widely used for human food and livestock feeds, starch production and chemical industries, while the straw is used for roofing huts and animal bedding. Grazing is sometimes performed after harvesting or when the crop is green [4, 7].

There are constraints facing barley-producing nations such as Australia; including transient, unpredictable and varying climatic conditions [8–10]. These environments are characterized by a lack of adequate water in spring and summer periods when evaporation and transpiration are rising rapidly when crops are in the later stages of development, which results in a terminal drought. There is also a problem of frost when the air temperature drops to 2°C or less. Damage to crops from frost may occur at any stage of development but is most damaging at and around flowering. These constraints result in a serious dilemma for growers who must decide whether to delay anthesis to avoid frost damage or flower as early as possible in order to escape the effects of terminal drought [11]. Thus, it is important that barley cultivars demonstrate an adaptation with appropriate rates of development across the heterogeneous environments.

1.1. Barley phenology—its relationship with abiotic stresses, quality and yield

Plant phenology characterises the developmental life cycle events of plants and how these events are influenced by seasonal and inter-annual variations in climate as well as habitat factors [12, 13]. In barley, different development stages, such as spikelet initiation and duration of grain development can seriously influence yield and quality. These stages are regulated by environmental factors such as temperature or growing degree days (GDD), duration and intensity of light, nutrition and husbandry techniques [14]. Heading date is important in adapting barley genotypes to different stresses such as heat stress, waterlogging, salinity and drought. Heat stress can quickly deplete the available moisture through high rates of evapotranspiration and ultimately leading to terminal drought [15]. Both heat and drought at late sowing may interrupt barley developmental processes usually from double ridge (DR) to maturity. The resultant effects of these stresses are reduction in plant height, dry matter accumulation and grain yield [15]. On the other hand, low temperature at early growth stages (Zadoks GS10) may be required for vernalization especially for winter barleys to flower. Apart from the optimum conditions, poor biomass accumulation and significant yield losses are attributed to extreme conditions, high or low temperatures, drought or waterlogged anaerobiosis and other soil-related problems [16]. Advances in crop phenology and modelling have helped with the understanding of how to assess biomass partitioning and effects of abiotic stresses in crops [12]. Modelling has also helped understand the effects of different environments and sowing dates on growth and development of barley plants.

Many scoring systems for plant growth stages have been developed to describe phenology in cereals [14]. The most widely used scales are Feekes scale [14] and Zadoks scale [17]. The majority of the scales described only the morphological traits [12], while very few describe the apical developmental processes especially on barley [18]. The vital developmental stages that have significant effects on yield and quality are DR (Zadoks GS30), construction phase which includes stem elongation (31), heading and anthesis (51&61 as in barley) and grain filling stage [19, 20]. DR and terminal spikelet (TS) can only be detected through destructive examination (**Figure 1**). There have been no consistent reports on the use of correlated traits for the determination of the flower/floral initiation stage (GS30), although in some cases surrogate traits such as number of leaves on the main stem have been used to determine this stage [21].

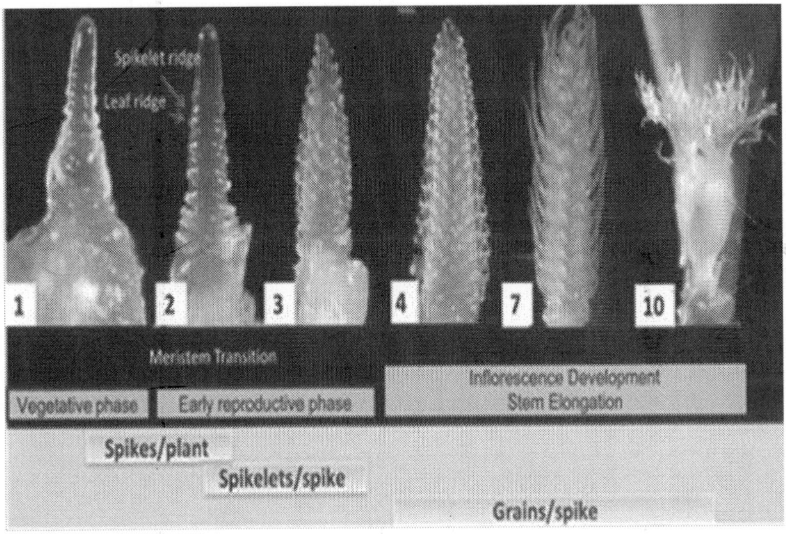

Figure 1. Development stages from double ridge up to terminal spikelet [151].

Developmental stages drive appearance of the main traits in cereals and are sensitive to climate and management [22, 23]. DR stage is an elementary step for predicting flowering date and crop yields and has been an important trait for improving crop productivity and adaptation [24]. Short construction phase (e.g. from Zadoks GS31 to 65) or short grain filling periods often lead to low yields [25]. The increase in the rate of grain filling is a positively correlated with grain weight [26]. A significant increase in grain yield in the high rainfall zones of Australia (usually along the coast with over 550 mm rainfall) was shown to be due to a longer duration between result from an increased time from GS31 to GS65 (i.e. stem elongation to anthesis) in wheat [27].

1.2. Factors affecting phenological development in barley

Barley development has three main stages: germination/emergence to double ridge/stem elongation (GS10 to GS30/31), stem elongation to heading/anthesis (GS30/31 to GS51/61) and heading to physiological maturity (GS51 to GS 94) [15, 28, 29]. The first stage, the basic vegetative phase, is the stage for the production of phyllochrons, roots and tillers. The agronomic significance of this stage is the generation of enough biomass for livestock feed especially for dual-purpose genotypes [30]. The second stage involves the termination of vegetative growth at GS29. This stage signifies reproductive growth, spikelet initiation and the onset of stem elongation [31] and is required for the production of a higher number of spikelets which directly link to grain yield [32]. The third stage is grain filling, influencing grain size and weight [33]. This stage is essential for yield increase as well as quality. All the three stages are regulated mainly by genetic, environment as well as management.

Important environmental factors affecting barley developmental stages include temperatures and photoperiod [15, 34, 35], both of which vary simultaneously in field conditions. This variability affects developmental events that determine flowering time and consequently yield [35]. Temperature is very important for all plant physiological processes [36], especially for the variation in days to spikelet initiation [37], days to heading and days to flowering (anthesis) in cereals [38, 39]. This observation is supported by Hay et al. [40] and Ellis et al. [41] who reported that the rate of primordia initiation especially spikelet in barley has a linear relationship with average daily temperatures. The rate of initiation of organs in barley also increases linearly with temperature; the optimum temperature for organ development is 25–30°C [37]. However, low temperature is required in some cereals to stimulate flowering (vernalization); this term has **been** used as the basis for classification of barley into winter and spring types. The variation in the development of phyllochrons in different genotypes of barley is more likely due to the variation in the combinations of temperature and photoperiod [12].

Growing degree days (GDD) are often used for measuring developmental events in barley. GDD is the accumulation of the mean daily temperature above a base temperature (0°C in the case of barley) [34, 42]. Below 0°C, the development of the crop will cease while above 0°C the growth will increase linearly with temperature [34, 36, 43]. Using zero as the base temperature (Tb = 0°C) in wheat, Acuna et al. [27] identified an ideotype which may develop sufficient tiller numbers at 650°Cd to harvest about 400–500 heads/m2. The same ideotype had a construction phase duration (CPD) within 800–1200°Cd to be able to escape frost and partitioned more assimilate to developing grain. Miralles et al. [35] reported the GDD that ranges between 950–2000°Cd and 1300–2100°Cd from sowing to flowering in both barley and wheat, respectively [35]. This result reflects the high variability in the GDD for flowering time in barley, indicating that yield might be manipulated using this characteristics.

Photoperiod is also a key environmental factor that affects the development of barley especially in temperate countries. The duration of day-length has a predictable pattern that drives evolutionary plant responses. This forms the basis of classifying barley as a long day, short day and day-neutral (DN) plant [44]. Photoperiod can significantly influence the duration of both vegetative spikelet initiation and stem elongation periods in wheat and barley [45]. The long day-length of higher northern and lower southern latitudes causes both photoperiod-sensitive and photoperiod-insensitive cultivars to flower early. The winter type responds strongly to long days, while spring types vary depending on the selection criterion [46]. For example, in Western Europe and parts of North America, short days increase the duration of vegetative growth of spring-sown barley. This lengthens the time for biomass accumulation and hence increases yield [46] under temperate growing conditions. Australia has a unique environment that differs from other barley growing regions in higher latitudes [21], where daylength is much shorter in winter, but considerably longer in summer (**Table 1**). However, there is variation in the extent of sensitivity of Australian genotypes to photoperiod. The baseline for the sensitivity ranges from 8 to 10 h of exposure, below which no flowering initiation occurs, while the upper limit ranges from 13 to 18 h [38]. The sensitivity of both vernalization and photoperiod starts immediately after plant emergence [38].

Region	**Coordinate**	**Seasonal sunshine hour duration**			
		Winter: June–August	**Spring: September–November**	**Summer: December–February**	**Autumn: March–May**
NT, Darwin					
NT, Katherine	14.465°S, 132.26° E	11–12	12–13	12–13	11–12
Perth	31.95°S, 115.87°E	10–12	11–14	12–14	10–13
Carnarvon	24.88°S, 113.66°E	10–12	11–14	12–14	10–12
Victoria, Mallee	35.11°S, 142.36°E	9–11	11–14	12–14	10–12
Ravensthorpe, WA	33.32°S, 119.82°E	9–11	11–14	12–14	10–12
SA, Wimmera	37.82°S, 140.78°E	9–11	11–15	13–14	10–12
Adelaide	34.93°S, 138.60°E	9–11	11–14	13–15	10–13
Victoria	37.47°S, 144.79°E	9–11	11–15	12–15	9–13
Brisbane, NSW	27.47°S,153.02° E	10–12	11–15	12–14	10–13
Tasmania	42.88°S, 147.32°E	9–10	11–15	13–15	9–13
NSW, Queensland	20.92°S, 142.70° E	9–12	11–14	12–14	12–14
Launceston, Tas	41.44°S, 147.14°E	9–11	11–15	13–15	9–13

Table 1. Sunshine duration across Australian regions and their coordinates [152].

Management is also an important factor that affects phenology in barley; important factors include sowing date, fertilizer application, irrigation and other management practices. Matching the phenology with an appropriate sowing window allows growers to better manage climate risks that are particularly pronounced in Australia, where early sowing may expose the heading of spring barley to frost, while late flowering and terminal drought can curtail grain filling and hence reduce yield [11]. Plant growth and development are affected by high temperatures and water stress in late sowing of spring barley [15, 47], high temperatures at early sowing in Punjab of India [47] and low temperatures with early sowing in Russia [15]. In the same experiment with early sown crops, low tillering capacity as a result of low temperatures caused a significant reduction in grain yield [15]. Similar results were reported by Ram et al. [47], in that very early sowing of spring barley may affect tillering capacity, although in this case due to high temperatures and reduction in biomass accumulation for late sowing. In addition, sowing date was found to have the most significant effects on the phenological development of cereals, especially during the GS31 (stem elongation phase) in wheat [48].

2. Physiological and molecular mechanisms for heading date and their effects on grain yield and quality

Heading date (spikelet emergence, Zadoks GS51) is a complex trait in barley that has direct impact on grain yield and quality and also forms the basis of evolutional adaptation to the changing climate. The mechanisms of regulating this trait are so complicated that there has been no final conclusion on the specific number of genes that are involved and their interactions [37]. Although the expression of this trait is governed by complex factors such as genetics, physiology and environment [21], neither the plant breeder nor the physiologist can clearly explain their interactions. For example, physiologists have not answered some unresolved developmental issues such as the regulation of the developmental rate within an environment and the cause of the transition between one growth stage to another. Equally, crop breeders need to account for the gene functions, the number of the genes and their interactions that are involved in expression of heading date [37].

2.1. Genetic regulation of heading (flowering) date

Barley improvement dates back to its domestication period. However, significant yield improvements began only in the 1950s as a result of the application of more advanced plant gene technologies [49]. Studies have been conducted on the genetic improvement, which include the determination of genotypic and phenotypic variability in days to flowering and growth phases, as well as comparing different sets of cultivars in barley [29, 50] and wheat [51]. Recent advances in the area of molecular breeding using highly polymorphic molecular makers such as simple sequence repeat (SSR) and single nucleotide polymorphisms (SNPs) have also led to significant progress in the improvement of yield and quality in barley. These markers are being used to tag genes or quantitative trait loci (QTLs) that are of economic importance, offering promises in their use in marker-assisted selection (MAS) [52]. Among all the markers, SNPs and SSRs are unanimously believed to be best suited for the use in marker-assisted breeding [53]. They have been used to assess most of the genes in barley and other cereals through cDNAs, expressed sequence tags (ESTs) and sequenced PCR amplicons that provide use of SNPs in protein encoding transcribed genes.

A genetic study on early versus late heading in barley was first reported in 1907 [54]. Exploration of the effects of environment on heading date was initiated following the reports of Garner et al. [55]. This opinion was supported by other studies that sources of variation in flowering date among different genotypes of barley were due to the effects of seasonal variation, location and sowing date [21]. As a result, genotypic differential response to photoperiod, vernalization and other environmental conditions have been conducted. Three groups of genes are responsible for variations in heading date. These include Ppd (photoperiod) [20, 44, 56, 57], Vrn (vernalization) [58, 59] and earliness *per se* (Eps) [60, 61]. Eps determines the time and duration of the reproductive phase. Of the three genes, only Eps gene acts independently of vernalization and photoperiod [19, 20, 62, 63]. All the Vrn genes have been cloned [46, 64–66]. All the three groups of genes have been well researched in wheat, although there is no conclusive evidence that Eps genes have been cloned in all cereals. These genes start to regulate development from emergence GS10 [20]. Hence, there is a lack of information on the genetics, physiological, biochemical functions and location of the Eps gene especially in barley [64, 67, 68]. Unfortunately, the access to the barley genome has not been straightforward because the genome consists of a large number of repetitive sequences [63, 69]. Scientists have explored the opportunities in the colinearity of the genes among the cereals [63, 67, 70, 71] for marker design and elucidation of the effects of the genes on yield and quality and their interaction with the environment in barley [63]. It is, however, important to understand the behaviour and interaction of these Eps genes with different environment and management practices. Particularly in Australia, variable climates make production decisions and genetic improvement for crop adaptation difficult [9]. Evidence of deferential genotype responses to ambient temperature and other climatic parameters is limited in barley [21]; thus, the knowledge of genotype x environment or QTL × environment as well as management (G × E × M) interactions is required to help obtain higher grain yields and quality [72]. The use of crop simulation modelling to predict expression of complex crop traits under diverse environments has provided plant breeders and farm managers with good opportunities to make crucial decisions such as matching the choice of genotypes to an appropriate sowing window or soil type, in different environmental conditions [73]. Therefore, integration of experimentally determined genetic responses to photoperiod, vernalization and Eps will complement plant breeders in their use of genetics and molecular tools in the prediction of flowering time and the understanding of how these genes affect grain yield and quality in different climates and with different management.

2.1.1. Photoperiod genes (Ppd)

The photoperiod pathway is generally classified into two components: the circadian clock and the photoperiod clock regulators [74]. The clock is the receptor of light stimuli perceived by phytochromes (phyA to phyE) and cryptochromes (cry1 and cry2) which are red and far red receptors and blue light receptors, respectively [74]. Temperate cereals, including barley, are quantitative long-day crops [21]; however, some varieties differ in their response to photoperiod [75], as mentioned above. Photoperiod sensitivity can significantly influence the duration of both vegetative and stem elongation periods in wheat and barley [45]. Longer day-length causes both photoperiod-sensitive and photoperiod-insensitive barley cultivars to flower early. There is evidence that specific stages in floral development in wheat and barley may also be sensitive to light intensity [76] and heavy shading during the later stages of ear development may result in the infertility of the spike [76]. As a result, flowering in photosensitive plants like barley may be entirely inhibited if the light intensity is reduced sufficiently during long periods because of the low level of availability of carbohydrates within shaded plants [76]. Therefore, genotypes vary in the photoperiodic threshold below in which flowering initiation will not take place.

In barley, two photoperiod genes influencing flowering time are *Ppd*-H1 on chromosome 2(2HS), which regulates flowering time under long days [56, 57, 62, 77, 78] and *Ppd*- H2 on chromosome 5(1HL) that regulates flowering time under short days [20, 62, 78]. *Ppd*- H1 is a pseudo-response regulator gene (HvPRR37) [79] and is the major controller of heading date when crops are exposed to long days (LDs). Therefore, the spring varieties of barley consist of this dominant allele. However, the recessive allele ppd-H1 is the major causes of the reduction in photoperiod response in European spring types and hence the reason for late flowering in LDs [46]. Reduced photoperiod responsiveness of the ppd-H1 mutant, which is highly variable in long season conditions, is explained by altered circadian expression of the photoperiod pathway gene CONSTANS and reduced expression of its downstream target, HvFT1, which is controlled by HvCO1, a key regulator of flowering [64, 80]. EARLY FLOWERING3 (ELF3), which is also a member of the circadian clock genes, regulates flowering under the influence of photoperiod [81]. This gene also has a loss-of-function mutant in plants (e.g. barley and some legumes) that causes early flowering in short days (SDs) as well as in LDs. In the same way as the ppd-H1 operates, the recessive mutant *eam8* (mat-a) has a loss of function characteristic [64] that leads to the insensitivity to photoperiod and thus can cause early flowering in both SDs and LDs [64, 81]. However, *eam8* is significantly involved in the expression of *HvFT1* (a flower initiator) which is also an allelic variant at Ppd-*H1* locus [64]. Similarly, the barley *elf3* mutant helps in the expression of the GA20oxidase2 gene, which causes the production of gibberellin (GA) in the apical meristems under SDs. Thus, the production of GA activates the early-flowering *elf3* in SDs in the absence of the *FT1* gene [81]. The second photoperiod gene (*Ppd-H2*) responds to short day-length. *Ppd-H2* acts similarly to *HvFT3* when exposed to SDs. In an experiment conducted using Morex and Steptoe populations, the expression of the HvFT3 was not found in the Steptoe genotype (which has the ppd-H2) but was found in in Morex (which has the Ppd-H2 gene). Therefore, *HvFT3* has been named as the candidate for *Ppd-H2* [82]. In spring barley, the Ppd-H2 allele is the major actor regulating flowering, but is rarely found in commercial winter types [82].

Many other QTLs have been identified from different populations. Ren et al. [57] also detected a major QTL under 18-h photoperiod in glasshouse experiments and mapped the QTL to the Xp12m50B199–Xp13m47B399 interval of flanking markers on chromosome 4H which accounted for 77.48 and 37.81% of phenotypic variation for long photoperiod response in Australia and China, respectively. The gene, *eam7*, showed a stronger effect on flowering time with 55 day and 18 day differences compared to Ppd-H1 (chromosome 2H) and Ppd-H2 (chromosome 1H) [83]. Another *eam7* determining photoperiod insensitivity under short day- length was identified on the short arm of chromosome 6H near the centromere [83]. This gene was 6.7 and 13.0 cM away from two flanking markers Xmwg2264 and Xmwg916, respectively. Environmental factors also had a significant effect on the expression of two different QTLs, for flowering time which were mapped to chromosomes 1HL and 7HS when the population was grown under long photoperiod conditions. However, no QTL was detected in the same lines when they were grown under short photoperiod conditions [78]. The QTL for heading date are often linked with yield in barley [19, 84]. These could be the part of the reasons why most of these genes have a highly significant effect on several agronomic traits, such as biomass accumulation including grain yield and grain quality in barley [20, 85]. The photoperiod responsive genes in wheat were found to be in homoeologous series to genes on barley chromosome 1H, 2H [28].

2.1.2. The vernalization genes (VRN)

Vernalization is the requirement for prolonged low temperature to advance flowering in cereals and depends on the growth habit (such as spring or winter types). The winter types of barley require cold exposure before flower initiation, typically below 10°C for a period between 4 and 6 weeks [86], depending on base temperature as defined above. In contrast, the spring types have minimal low temperature dependency and are usually insensitive to vernalization and Short day photoperiod [87]. This behaviour is characteristic of many temperate cereals like barley [87, 88] and associated with the capacity of a genotype to survive the cold winter during vegetative stages [87, 89, 90]. Barley is an excellent model for genetic analysis of low-temperature tolerance in fall-sown cereals [90]. Its responses to vernalization have been observed to vary greatly among genotypes and between growth phases [45, 90].

Vernalization in cultivated barley is mainly controlled by three major Vrn genes [28], Vrn1, Vrn2 and Vrn3 [58], or HvVrn1, HvVrn2, HvVrn3 [88], or Vrn-H1, Vrn-H2 and Vrn-H3 [59]. The Vrn-H1 (also named as Sgh2 or Sh2) is located in the middle of the long arms of 5H [67, 88]. The Vrn-H2 (Sgh or Sh) is found on chromosome 4H [67], while the Vrn-H3 (Sgh3 or Sh3) is on 1H [59]. Vrn-H1 translates the fruit-like MADS-box transcription factor which is an ortholog *APETA-LA1* gene [65]. The allelic difference at this gene locus is essential for flowering in temperate cereals [65, 88, 91] and therefore, it is one of the major determinants of vernalization requirement in barley and wheat [92]. Within the locus, the allele that is responsible for the spring growth habit is Vrn-H1 (the dominant one) [93], while the recessive allele accounts for genetic regulation of the winter habit [65, 67, 88]. A large deletion in the first intron of Vrn-H1 locus in the dominant allele is responsible for the null response to vernalization in spring barley and wheat [93, 94], while no deletion within intron 1 was observed in the winter habit types possessing recessive vrn-H1 allelic loci [67, 88, 93].

The second locus is the Vrn-H2 (Sgh2 or Sh2) which encodes for the zinc finger-CCT (ZCCT-H) transcription factor [93] and is also vernalization dependent. A partial or total deletion of part of this locus has been shown to cause a non-functional mutation of the gene and a recessive form is responsible for the spring growth habit in both barley and wheat [65, 92, 95]. However, it is necessary to understand that the effects of Vrn-H2 under field conditions can only be verified using a variety of sowing dates [59]. The authors further stated that the gene does not affect heading date when crops were autumn sown.

The third is the Vrn-H3 (Sgh3 or *Sh3*) on chromosome 1H [20, 28, 88] and later on 7HS [62, 96]. This gene is an ortholog of the FT gene in *Arabidopsis* [96] and HvFT1 gene [62] which responds to vernalization in both barley and wheat. A study conducted by [96] showed that homologous spring barley with dominant Vrn-H3 allele had an increase in

HvFT transcript rapidly, while the recessive genotype vrn-H3 had low HvFT transcript without vernalization. A strong relationship was found between the Vrn-H3 and Ppd genes as the HvFT was observed to be very low in SD and upregulated in LDs [96]. Finally, for a given winter genotype to respond to vernalization, it must have all three (vrn-H1:Vrn-H2:vrnH3) and all other combinations are reported to be spring types [59, 88, 95]. These three loci (VRN-H1, VRN-H2 and VRN-H3) interact in an epistatic fashion to determine vernalization sensitivity [95].

Since there is a form of homogeneous genetic system for all the cereals with a high degree of synteny (physical co-localization of genetic loci on the same chromosome in an individual or species), the results of one species are frequently applicable to other members of the cereal family [87, 97, 98], including barley. Consequently, the cloning of the candidate genes in diploid wheat (*Triticum monococcum*) of VRN-Am1 and VRN-Am2 [65, 91, 92, 99] and hexaploid wheat (T. *aestivum*) of VRN-1 has considerably increased our understanding of the genetics of vernalization in barley [87].

2.1.3. *Basic vegetative phase BVP (Earliness per se, Eps)*

Barley has the potential to grow and produce economically viable yields under a wide range of diverse environment-types. Early growth plasticity is determined during the vegetative phase [100], which has extensive genetic diversity. One of these genes is the *Earliness per se*, Eps. This gene regulates the basic vegetative phase (BVP) in barley and influences the time and duration of growth stages from DR (Zadoks GS30) to grain filling stage (Zadoks GS70) [60, 61]. Expression of this gene can only be fully observed when all the other sources of the variations in flowering time have been fixed, i.e. when the environmental stimuli such as exposure to adequate vernalization and photoperiod requirements have been met by the plants [70, 71, 101–103]. In addition, Eps is also actively involved in the fine-tuning of the flowering time in cereals including barley [71, 104]. Various authors have identified the Eps gene in all the chromosomes of common wheat [71] and barley [20, 63, 67]. Recent advances in molecular genetics have shown that the location and physiological effects of the Eps gene on yield and quality in barley are limited [63]. Since most of the cereals share similar genetic synteny [63], it could be assumed that results from studies on reports on wheat could be applied to barley. Efforts to identify the markers linked to the genes and their locations are underway. In wheat, RFLP marker, wg241 was observed to be linked to Eps-Am1 gene on 1H [71]. The gene was found to be 0.7 cM distal to wg241 and 1.4 cM proximal to the barc287 markers [60, 71]. Among the three markers reported in *Brachypodium* and wheat plants, two were identified to be molybdenum transporter 1 (Mot1) (transcriptional regulator) and filamentation temperature-sensitive H4 (FtsH4), respectively [70, 105]. These makers were linked to the Eps gene and were proposed as candidates for Eps-Am1 on chromosome 1H [70, 105]. The predicted MOT1 protein showed differences in the amino acid between the parent lines in which the effects could not be predicted [105]. Thus, any future steps to clone the Eps-Am1 gene should include the generation of mot1 and ftsh4 mutants and the completion of the T. *monococcum* physical map to test for the presence of additional candidate genes.

2.2. Effects of Eps gene on developmental phases

The genetic and physiological processes that are linked to the adaptation of barley are due to broad differences in the developmental phases. These phases include both the time and duration from spikelet initiation (GS30) and up to grain filling (Zadoks GS70). Most previous research conducted has centred on the effects of Eps gene on the flowering time [61, 65, 67] with fewer focussing on the variations in the duration of each of developmental stages [61, 106, 107]. For example, an study conducted by Lewis et al. [61] using single seed decent and near isogenic lines (NILs) observed a significant interaction between the Eps gene and the timing and duration from vegetative to flowering phase especially from double ridge to terminal spikelet stage in diploid wheat. The interaction showed that the NIL genotypes with the early allele, Eps-e, had the transition to DR stage 35 days earlier (67% less) than the genotypes with the Eps-l alleles. The SSD genotypes had highly significant differences ($P < 0.0001$) in both heading time and number of spikelets per spike between Eps alleles (eps-e and eps-l). The genotypes with the late allele Eps-l flowered 61 days later than those with eps-e alleles (with 76% across temperatures) and produced a mean of 8.7 more spikelets for each spike which was a 56% increase across temperatures [61]. However, results of Valárik et al. [71] and Zikhali et al. [103] showed only a few days of differences (from 1 to 5 days) in flowering time between a pair of near isolines (NILs) and their recombinant inbred lines (RL) in both wheat and rice. In addition, no significant interaction ($P = 0.67$) was observed between Eps genes and the stem elongation stage [61], which is the beginning of construction phase. Also temperature had no significant effects on the gene determining spikelets number per spike [61]. Contrary to this opinion, Slafer et al. [108] observed that lengthening the duration of the stem elongation phase, without modifying total time to anthesis, could increase the number of grains/m2 and consequently the number of grains per unit land area [45]. In general, variations in both flowering time and spikelet number per spike could be due to pleiotropic effects of a single gene or to the effect of tightly linked multiple genes with additive effects [71, 105].

2.3. Effects of temperature on the Eps genes

Temperature is the major environmental factor affecting Eps genes in barley and wheat [61, 101]. Differences exist among genotypes carrying Eps-e for early heading and Eps-l alleles for late heading in wheat [60]. There are significant interactions between the Eps gene and temperature [60, 61]. The genotypes with Eps-l alleles had no interaction with temperatures (21°Cd difference), while lines carrying the Eps-e allele had a shorter thermal time to heading at 16°C than at 23°C (336°Cd difference) [61]. Lewis et al. [61] further showed that the thermal time to flowering for the genotypes with the Eps-e gene was approximately 1557°Cd. These are 1118°Cd less than the thermal time for the late genotypes (2675°Cd) with Eps-l gene. Slafer et al. [106] used four wheat varieties and six differential temperatures (10–25°C) to study the effect of temperature on growth stages. They showed that the developmental phases of individual genotypes were most sensitive to temperature from sowing to anthesis. This variation can be attributed to the allelic diversity at Eps locus in the lines studied. Hence, an in-depth research of genetic variability of the earliness *per se* genes (Eps) is required for a more precise analysis of their effects on developmental stages and temperature sensitivity.

2.4. Effect of sowing date on the Eps genes

In order to maximize yield potential in any environment, cultivars must have an appropriate flowering window and life cycle duration in the target environment [28, 60]. Sowing date is an important factor that governs flowering period, the timing of which needs to escape biotic and abiotic stress. Out of these three major genes, Ppd, Vrn and Eps, Eps genes do not respond to the differential to vernalization or photoperiod and still control timing and duration of flowering independent of these stimuli [28].

2.5. Effect of the Eps gene on grain quality

Yield and quality are important complex traits in any breeding programme. Improvement of these traits is very difficult to achieve due to their genetic, physiological and physical complexity. Grain quality could either be physical, such as size, hardness and lustre, or nutritional such as malting quality. The relationship of Eps gene on heading, spikelets number and number of grains per spike has been shown to be correlated with yield [61]. However, grain quality can also be improved by manipulating the Eps gene loci [109] to improve the physical traits such grain weight or hardness. Experiments conducted by Herndl et al. [110] indicated that with a shorter pre-anthesis period, the relationship between yield and protein is always negative. Crop breeders often focus on increasing yield with little attention on quality traits; thus, there is less information on the effects of the Eps gene on quality traits.

Heading date is a polygenic trait, controlled by Ppd [111, 112], vernalization [37, 38, 41, 111] and Eps genes [111, 113]. These genes interact in an additive nature (cumulative effects of non-allelic genes to a quantitative trait) [111]. The genetic analysis of Eps showed that it can be simply inherited as a Mendelian inheritance, but molecular analysis has not been able to identify an appropriate molecular marker to determine its location [114].

3. Modelling

Crop modelling in agriculture has been used as a physiological framework to undertake simulation of dynamic crop phenology that support crop improvement programmes [115]. Physiologically sound simulation tools will provide quantitative assessments of crop development and yield relative to the genotype, climate, soil and management in sustainable farming systems [116]. These tools should provide ex-ante impact assessments of research outcomes across a wide range of environments [100]. This is particularly true for Australia where a highly variable climate poses challenges for production and crop improvement [9]. With respect to climate change, temperature has increased by 0.9 °C since 1910 per annum and severe heat and drought spells are occurring more frequently in Southern Australia. Total annual rainfall and frost events may also increase in some of the temperate regions such as Tasmania [117]. Other reports indicate a 2–5% reduction in rainfall across most parts of the country [118]. Harrison et al. [119] emphasizes the need for serious attention on the impacts of climate stress on plant phenology. Globally, climate change will further increase temperatures, modify the amount and distribution of rainfall and consequently reduce the probability of reliable food and forage production, thereby causing a significant threat to food security and improved livelihood [120]. More than 30% barley yield loss will be attributed to climate change as a result of drought and heat stress by 2050 [121]. Despite this, there has been little research work to determine how the climate change will broadly affect whole farm systems [119, 122], including systems farming barley.

Mathematical functions are being used as tools to simulate crop phenology to predict the effects of climate events and changing environments on yield and quality [24]. These tools help explain the interaction of some of the complex traits related to development and growth and their interaction with the environment. For example, ecophysiological quantitative equations were used by Yin et al. [24] to describe the response of flowering to photoperiod and temperature to predict days to heading and yield in diverse conditions. The equations were used as empirical and mechanistic models to provide important framework for simulating a number of events in crop growth, especially predicting heading date [24, 123]. The empirical models are often based on accumulation of growth degree days adjusted by vernalization and photoperiod [123, 124], while mechanistic models are based on the production of leaves and floral primordia at the apexes [125].

Four important phenology models: 3s-beta-model, 3-plane-linear-model, modified-rice-clock-model (m-RCM) and a logistic model were developed and evaluated in rice [39, 126]. All models were able to predict the flowering time in varying environments although with varying degree of precisions. Model parameter values from reciprocal transfer experiments also resulted in realistic differences in flowering time across all the genotypes in different environments. The models were able to partition variation due to environment and that of the genotypes. Thus, ecophysiological model could be very important for dissecting the relationship between genotype and phenotype [24]. Chapman et al. [73] also developed mechanistic model called QSUN to estimate growth, development and yield of a diverse range of genotypes of sunflower under varied environments. Their model was able to account for leaf area index ($r2 = 0.65$), total biomass ($r2 = 0.96$) and grain yield ($r2 = 0.93$) when tested against actual phenological data. QSUN was also used to analyse the production risk of sunflower grown in highly variable subtropical environments in order to undertake decisions such as the choice of an adapted cultivar and appropriate sowing window in order to obtain higher yields [9]. Another dynamic model was used to investigate the causes and impact of climate change on peanut production in Northern Australia [127]. The model was used in conjunction with the information of district yield to offer an in-depth study of long- term production risk. The study indicated that the stabilization of the above-average yield, which was due to stable summer rainfalls, was responsible for the rapid expansion of peanut industries at that time. Such studies assist in gaining better understanding of complex GxExM and the identification of traits required to manage crops in variable and changing environments [127]. Therefore, choice of an appropriate simulation models for predicting phenology are essential for choosing the best-adapted cultivar for a specific production environment and for helping with timely planning of strategic or tactical management [9, 123].

3.1. The agricultural production systems simulator (APSIM)

The Agricultural Production Systems Simulator (APSIM) is a cropping systems simulation model that combines several decision-support tools. APSIM may be used for accurate predictions of how traits like heading dates impact on grain yield and biomass of different crop genotypes in alternative environment and management conditions and also to consider the long-term consequences of cropping systems on soil conditions [116]. APSIM may also be used to increase the understanding factors influencing heading date of barley when grown under field conditions [12, 115]. The tools within APSIM can also be used to describe genetic parameters regulating phenology with the function of daily temperature and photoperiod to predict flowering time and consequently yield and quality [126]. The major challenges facing barley production are water stress coupled with heat stress during spring and summer and frost events in winter and early spring. District yield records showed about 85% yield loss due to frost events in Australia [128]. A later study with more information was conducted to gauge the impact of frost on grain production in Australia [129]. In the study, APSIM simulated the effects of frost on wheat production areas across Australia. The model predicted increased frequency of frost events in the Australian wheat belt (the main barley production regions in Australia) and also an increase in the mean temperatures with significant yield loss. Zheng et al. [129] concluded that breeding for frost tolerance could give about 20% yield advantage. As for most other modules, the barley module of APSIM (APSIM-Barley) simulates the phenology in a daily time step. The module uses inputs of weather such as radiation and temperature and initial soil nitrogen [130]. The module has 11 growth stages, from sowing to harvest (GS0–GS100) [130]. Manschadi et al. [131] took advantage of APSIM's scientific basis by assessing barley growing patterns under different environment and management. The model was able to account 91% and 82% of the variation for biomass accumulation at maturity and grain yield. Although negative correlation between yield and quality traits [132] and between life cycle and quality traits [133] were reported, maintenance of quality traits is critical in order to alleviate malnutrition [134]. The majority of crop models are constrained in predicting both the physical and cryptic (nutritional) grain quality such as grain size and grain-N content [134], although the model has been used to account for both above and below ground biomass, growth, water, N uptake and leaching [124]. The same model was used to explain some quality parameters such as grain size and grain protein concentration [135]. As a result of the challenges due to frost and drought events faced by barley production, a simulation model, QBAR, was developed to identify appropriate management options such as sowing date in order to increase yield [136]. Thus, QBAR can simulate phenology, soil water, leaf area, biomass accumulation and yield of barley [136]. A more detailed analysis of extreme terminal drought effects and frost risk should be conducted that should include several sowing dates and varieties. QBAR was later modified to (APSIM-Barley) [136], accounted for 91 and 82% variances for biomass accumulation and grain yield, respectively [137]. Further improvement on the QBAR model has been the integration of crop nitrogen balance and grain quality module [136]. It was later used to account for the effects of extreme climatic events, frost and terminal drought on yield and yield components, of which paddock-based crop models could not explain [138]. The authors proposed that QBAR can be used to determine the best management decisions such as sowing date to obtain highest grain yield even in the events of frost and water stress.

3.2. Genotype by environment by management (G × E × M)

Information on the target environment for which crop cultivars are to be improved is vital to plant breeders [139]. This is because a higher genetic (G) diversity for flowering time has been reported in diverse environments (E) of barley growing regions where frost and drought events limit crop growth as well as inappropriate agronomic management (M) to improve crop growth and yield [72]. This concept can be extended to include both abiotic factors such as soil, water stress and waterlogging as well as induced stresses due to abiotic (salinity) and biotic factors (pests, weeds and diseases). Environmental factors can be classified into two types: (1) micro-environmental factors such as year-to-year variation in rainfall, drought conditions, pest incidence and (2) macro-environmental factors which include soil type and management practices [140, 141]. The association between the environment and the genotype to produce a specific phenotype is termed as the G × E interaction [141]. Hence, the G × E interaction determines the adaptability and suitability of a specific genotype to a range of environments. Environment could also be a time boundary, such as a year (annuals) [142]. Hence, matching heading date to diverse environment may give a large G × E interaction [142]. Variation in the developmental stages usually from DR to grain filling stages in barley is influenced by G × E [143]. For highly variable environment types like those found in Australia, there is a need for specific adaptation (genotype response and better performance in a specific environment) arising from G × E interaction [144]. Löffler et al. [145] used a crop model index approach to account for the G × E interaction effect in US maize breeding trials. Factorial regression (FR) has been used to describe crop interaction with their environment and help understand G × E [146]. A linear generic model was used to analyze the interaction of 96 genotypes to different environment [126]. The model was able to explain 81% of the total variation in heading date across the environments. The introduction of molecular markers has aided our understanding of the effect of individual gene or QTL effects rather than the cultivar [126]. Yin et al. [126] used a four-parameter ecophysiological model to predict grain yield when QTL-based data inputs were used. The model when used together with the QTL map was able to sufficiently predict days to flowering in barley [126], suggesting that the model could be used to help breeders in Australia to adapt new varieties.

Recent advances in plant breeding combined with dynamic models are now allowing partitioning of the effect of management in the G × E approaches [72]. Higher genetic gain (GA) in yield has been attributed to better understanding of G × M interaction effects in maize crops, where the progressive yield increase in the US has been associated with superior genotypes being grown at higher density [147]. Another example is the choice of a combination of non-tillering genotypes (G) and row spacing (M) in drought prone land can help realize sustainable production and additional value in obtaining moderate yield instead of complete crop failure due to limited availability of water [72, 148]. Another study was conducted to check the performance chickpea genotypes under two different managements, irrigation and rain- fed management systems. The study showed highly significant yield differences among genotypes and between the two management practices for all the important traits. The study also revealed that both yield and yield components were improved by an average of 48% increase in the number of pods per plant, 36% in total dry weight and 17% in grain yield in the management involving irrigation [149].

It is therefore important to note that the use of models capable of accounting for G × E × M interaction in breeding and agricultural systems can be a powerful tool to better understand environment-specific, complex gene expressions. APSIM-Barley has been used to describe broad adaptation of barley genotypes in anticipation frost or water stress across Australia. In another experiment, leaf area and yield of Baudin, Flagship, Buloke and Capstan, barley cultivars were assessed. The model also reasonably explained the relationship between the leaf area duration and yield as influenced by weather [150]. However, the model has not been used to explore the potential agronomic benefits of exploiting G × M interaction in a specific environment. Matching specific genotypic traits to management option within a target (specific) environment will assist breeders in trait selection and the design of their breeding programs.

Author details

Ahmed Ibrahim[1,2], Matthew Harrison[1], Holger Meinke[1] and Meixue Zhou[1*]

*Address all correspondence to: Meixue.Zhou@utas.edu.au

1 Tasmania Institute of Agriculture, University of Tasmania, Hobart, TAS, Australia

2 Department of Plant Science, Ahmadu Bello University, Zaria, Nigeria

References

[1] IGC, Five-year global supply and demand projections. International Grain Council Report, http://www.igc.int/en/downloads/grainsupdate/IGC_5yearprojections2015.pdf, 2015.

[2] ABC, Barley crop trial for whisky production beats poor season to yield valuable data. Australian Broadcasting Corporation, 2015. http://www.abc.net.au/news/2015-02-05/whisky-barley-trial-crop-beats-poor-season-yields-valuable-data/6072166

[3] Barley Australia. Quality Barley for Australia and the world. 2015 [cited 2015 4 June 2015]; Available from: http://www.barleyaustralia.com.au/industry-information.

[4] CGIAR, Barley, 2015. http://www.cgiar.org/our-strategy/crop-factsheets/barley/

[5] Saisho, D., M. Ishii, K. Hori, and K. Sato, Natural variation of barley vernalization requirements: implication of quantitative variation of winter growth habit as an adaptive trait in East Asia. Plant and Cell Physiology, 2011. 52(5): p. 775–784.

[6] Baik, B.-K. and S.E. Ullrich, Barley for food: characteristics, improvement, and renewed interest. Journal of Cereal Science, 2008. 48(2): p. 233–242.

[7] Akar, T., M. Avci, and F. Dusunceli, Barley: Post harvest operations. The Central Research Institute for Field Crops, Ankara, Turkey, 2004.

[8] Australia, 2015. http://www.barleyaustralia.com.au/industry-information.

[9] Meinke, H., G.L. Hammer, and S.C. Chapman, A sunflower simulation model: II. Simulating production risks in a variable sub-tropical environment. Agronomy Journal, 1993. 85(3): p. 735–742.

[10] Index Mundi, 2015. http://www.indexmundi.com/agriculture/?commodity=barley&graph=yield.

[11] Richards, R.A, Crop improvement for temperate Australia: Future opportunities. Field Crops Research, 1991. 26(2): p. 141–169.

[12] Juskiw, P., Y. Jame, and K. L, Phenological development of spring barley in a short season growing area. Agronomy Journal, 2001. 93: p. 370–379.

[13] Wilczek, A., L. Burghardt, A. Cobb, M. Cooper, S. Welch, and J. Schmitt, Genetic and physiological bases for phenological responses to current and predicted climates. Philosophical Transactions of the Royal Society B: Biological Sciences, 2010. 365(1555): p. 3129–3147.

[14] Landes, A. and J. Porter, Comparison of scales used for categorising the development of wheat, barley, rye and oats. Annals of Applied Biology, 1989. 115(2): p. 343–360.

[15] Hossain, A., J. Teixeira da Silva, M. Lozovskaya, V. Zvolinsky, and V. Mukhortov, High temperature combined with drought affect rainfed spring wheat and barley in south-eastern Russia: yield, relative performance and heat susceptibility index. Journal of Plant Breeding and Crop Science, 2012. 4(11): p. 184–196.

[16] Cattivelli, L., S. Ceccarelli, I. Romagosa, and M. Stanca, Abiotic stresses in barley: problems and solutions. Barley: improvement, production, and uses. Wiley, Harrisonburg, 2010: p. 282–306.

[17] Zadoks, J.C., T.T. Chang, and C.F. Konzak, A decimal code for the growth stages of cereals. Weed Research, 1974. 14(6): p. 415–421.

[18] Banerjee, S. and F. Weinhues, Comparative studies on the development of the spike in wheat, barley and rye. Pflanzenzuchtung, 1965. 54: p. 130–142.

[19] Bezant, J., D. Laurie, N. Pratchett, J. Chojecki, and M. Kearsey, Marker regression mapping of QTL controlling flowering time and plant height in a spring barley (Hordeum vulgare L.) cross. Heredity, 1996. 77(1): p. 64–73.

[20] Laurie, D., N. Pratchett, J. Snape, and J. Bezant, RFLP mapping of five major genes and eight quantitative trait loci controlling flowering time in a winter× spring barley (Hordeum vulgare L.) cross. Genome, 1995. 38(3): p. 575–585.

[21] Boyd, W., C. Li, C. Grime, M. Cakir, S. Potipibool, L. Kaveeta, S. Men, M.J. Kamali, A. Barr, and D. Moody, Conventional and molecular genetic analysis of factors contributing to variation in the timing of heading among spring barley (Hordeum vulgare L.) genotypes grown over a mild winter growing season. Crop and Pasture Science, 2003. 54(12): p. 1277–1301.

[22] Cuesta-Marcos, A., A. Casas, P. Hayes, M. Gracia, J. Lasa, F. Ciudad, P. Codesal, J. Molina-Cano, and E. Igartua, Yield QTL affected by heading date in Mediterranean grown barley. Plant Breeding, 2009. 128(1): p. 46–53.

[23] Cuesta-Marcos, A., E. Igartua, P. Codesal, J.R. Russell, J.L. Molina-Cano, M. Moralejo, P. Szucs, M.P. Gracia, J.M. Lasa, and A.M. Casas, Heading date QTL in a spring× winter barley cross evaluated in Mediterranean environments. Molecular Breeding, 2008. 21(4): p. 455–471.

[24] Yin, X., P.C. Struik, J. Tang, C. Qi, and T. Liu, Model analysis of flowering phenology in recombinant inbred lines of barley. Journal of Experimental Botany, 2005. 56(413): p. 959–965.

[25] Schelling, K., K. Born, C. Weissteiner, and W. Kühbauch, Relationships between yield and quality parameters of malting barley (Hordeum vulgare L.) and phenological and meteorological data. Journal of Agronomy and Crop Science, 2003. 189(2): p. 113–122.

[26] Al-Karaki, G.N., Phenological development-yield relationships in durum wheat cultivars under late-season high-temperature stress in a semiarid environment. ISRN Agronormy, 2012 Article ID 456856.

[27] Acuna, T.B., Can the duration of the spike construction phase increase the yield of wheat? 2015.

[28] Snape, J., K. Butterworth, E. Whitechurch, and A. Worland, Waiting for fine times: genetics of flowering time in wheat. Euphytica, 2001. 119(1–2): p. 185–190.

[29] González, F.G., G.A. Slafer, and D.J. Miralles, Vernalization and photoperiod responses in wheat pre-flowering reproductive phases. Field Crops Research, 2002. 74(2): p. 183–195.

[30] Redmon, L.A., E.G. Kreazer, D.J. Bernardo, and G.W. Horn, Effect of wheat morphological stage at grazing termination on economic return. Agronomy Journal, 1996. 88(1): p. 94–97.

[31] Baker, C. and J. Gallagher, The development of winter wheat in the field. 2. The control of primordium initiation rate by temperature and photoperiod. The Journal of Agricultural Science, 1983. 101(02): p. 337–344.

[32] González, F.G., G.A. Slafer, and D.J. Miralles, Floret development and spike growth as affected by photoperiod during stem elongation in wheat. Field Crops Research, 2003. 81(1): p. 29–38.

[33] Whitechurch, E. and G. Slafer, Contrasting Ppd alleles in wheat: effects on sensitivity to photoperiod in different phases. Field Crops Research, 2002. 73(2): p. 95–105.

[34] McMaster, G. and W. Wilhelm, Phenological responses of wheat and barley to water and temperature: improving simulation models. The Journal of Agricultural Science, 2003. 141(02): p. 129–147.

[35] Miralles, D.J., B.C. Ferro, and G.A. Slafer, Developmental responses to sowing date in wheat, barley and rapeseed. Field Crops Research, 2001. 71(3): p. 211–223.

[36] Went, F., The effect of temperature on plant growth. Annual Review of Plant Physiology, 1953. 4(1): p. 347–362.

[37] Hay, R. and R. Ellis, The control of flowering in wheat and barley: what recent advances in molecular genetics can reveal. Annals of Botany, 1998. 82(5): p. 541–554.

[38] Roberts, E., R. Summerfield, J. Cooper, and R. Ellis, Environmental control of flowering in barley (Hordeum vulgare L.). I. Photoperiod limits to long-day responses, photoperiod-insensitive phases and effects of low-temperature and short-day vernalization. Annals of Botany, 1988. 62(2): p. 127–144.

[39] Yin, X., M.J. Kropff, T. Horie, H. Nakagawa, H.G. Centeno, D. Zhu, and J. Goudriaan, A model for photothermal responses of flowering in rice I. Model description and parameterization. Field Crops Research, 1997. 51(3): p. 189–200.

[40] Hay, R. and E. Kirby, Convergence and synchrony—a review of the coordination of development in wheat. Crop and Pasture Science, 1991. 42(5): p. 661–700.

[41] Ellis, R., E. Roberts, R. Summerfield, and J. Cooper, Environmental control of flowering in barley (Hordeum vulgare L). II. Rate of development as a function of temperature and photoperiod and its modification by low-temperature vernalization. Annals of Botany, 1988: p. 145–158.

[42] Bauer, A., AL. Black, AB. Frank, and E. Vasey, Agronomic Characteristics of Spring Barley in the Northern Great Plains. North Dakota State University Agricultural Experiment Station Bulletin, 1992. 523: p. 47. https://ndawn.ndsu.nodak.edu/help-barley-growing-degree-days.html

[43] McMaster, G.S. and W. Wilhelm, Growing degree-days: one equation, two interpretations. Agricultural and Forest Meteorology, 1997. 87(4): p. 291–300.

[44] Kikuchi, R. and H. Handa, Photoperiodic control of flowering in barley. Breeding Science, 2009. 59(5): p. 546–552.

[45] Miralles and R. Richards, Responses of leaf and tiller emergence and primordium initiation in wheat and barley to interchanged photoperiod. Annals of Botany, 2000. 85(5): p. 655–663.

[46] Turner, A., J. Beales, S. Faure, R. Dunford, and D. Laurie, The pseudo-response regulator Ppd-H1 provides adaptation to photoperiod in barley. Science 2005. 310: p. 1031–1034.

[47] Ram, H., B. Singh, and A. Sharma, Effect of time of sowing on the field performance of barley (Hordeum vulgare L.) in Punjab. Journal of Research, 2010. 47(3 and 4): p. 132–135.

[48] Kiss, I., K. Balla, J. Banyai, O. Veisz, and I. Karsai, Effect of different sowing times on the plant developmental parameters of wheat. Cereal Research Communications, 2015. 42(2): p. 239–251.

[49] Thomas, W., Prospects for molecular breeding of barley. Annals of Applied Biology, 2003. 142(1): p. 1–12.

[50] Borràs, G., I. Romagosa, F. van Eeuwijk, and G.A. Slafer, Genetic variability in duration of pre-heading phases and relationships with leaf appearance and tillering dynamics in a barley population. Field Crops Research, 2009. 113(2): p. 95–104.

[51] Whitechurch, E., G. Slafer, and D. Miralles, Variability in the duration of stem elongation in wheat and barley genotypes. Journal of Agronomy and Crop Science, 2007. 193(2): p. 138–145.

[52] Jahoor, A., L. Eriksen, and G. Backes, QTLs and genes for disease resistance in barley and wheat, in Cereal genomics. 2005, Springer, Kluwer Academic Publishers Dordrecht, The Netherlands. p. 199–251.

[53] Gupta, P.K. and R. Varshney, The development and use of microsatellite markers for genetic analysis and plant breeding with emphasis on bread wheat. Euphytica, 2000. 113(3): p. 163–185.

[54] Smith, L., Cytology and genetics of barley. The Botanical Review, 1951. 17(1): p. 1–51.

[55] Garner, W.W. and H.A. Allard, Effect of the relative length of day and night and other factors of the environment on growth and reproduction in plants. 1920. Journal of Agricultural Research Volume 18 Number 11, p. 553–606

[56] Wang, J., J. Yang, Q. Jia, J. Zhu, Y. Shang, W. Hua, and M. Zhou, A new QTL for plant height in barley (Hordeum vulgare L.) showing no negative effects on grain yield. PLoS One, 2014. 9(2): p. e90144.

[57] Ren, X., C. Li, M. Cakir, W. Zhang, C. Grime, X.-Q. Zhang, S. Broughton, D. Sun, and R. Lance, A quantitative trait locus for long photoperiod response

mapped on chromosome 4H in barley. Molecular Breeding, 2012. 30(2): p. 1121–1130.

[58] Distelfeld, A., C. Li, and J. Dubcovsky, Regulation of flowering in temperate cereals. Current Opinion in Plant Biology, 2009. 12(2): p. 178–184.

[59] Karsai, I., P. Szucs, K. Mészáros, T. Filichkina, P. Hayes, J. Skinner, L. Láng, and Z. Bedo, The Vrn-H2 locus is a major determinant of flowering time in a facultative× winter growth habit barley (Hordeum vulgare L.) mapping population. Theoretical and Applied Genetics, 2005. 110(8): p. 1458–1466.

[60] Bullrich, L., M. Appendino, G. Tranquilli, S. Lewis, and J. Dubcovsky, Mapping of a thermo-sensitive earliness per se gene on Triticum monococcum chromosome 1Am. Theoretical and Applied Genetics, 2002. 105(4): p. 585–593.

[61] Lewis, S., M.E. Faricelli, M.L. Appendino, M. Valárik, and J. Dubcovsky, The chromosome region including the earliness per se locus Eps-Am1 affects the duration of early developmental phases and spikelet number in diploid wheat. Journal of Experimental Botany, 2008. 59(13): p. 3595–3607.

[62] Wang, G., I. Schmalenbach, M. von Korff, J. Léon, B. Kilian, J. Rode, and K. Pillen, Association of barley photoperiod and vernalization genes with QTLs for flowering time and agronomic traits in a BC2DH population and a set of wild barley introgression lines. Theoretical and Applied Genetics, 2010. 120(8): p. 1559–1574.

[63] LÜ, R.-h., Y.-h. XU, R. Boyd, X.-q. ZHANG, S. Broughton, M. Jones, C.-d. LI, and Y.-f. CHEN, Barley and Wheat Share the Same Gene Controlling the Short Basic Vegetative Period. Journal of Integrative Agriculture, 2013. 12(10): p. 1703–1711.

[64] Faure, S., A.S. Turner, D. Gruszka, V. Christodoulou, S.J. Davis, M. von Korff, and D.A. Laurie, Mutation at the circadian clock gene EARLY MATURITY 8 adapts domesticated barley (Hordeum vulgare) to short growing seasons. Proceedings of the National Academy of Sciences, 2012. 109(21): p. 8328–8333.

[65] Yan, L., A. Loukoianov, G. Tranquilli, M. Helguera, T. Fahima, and J. Dubcovsky, Positional cloning of the wheat vernalization gene VRN1. Proceedings of the National Academy of Sciences, 2003. 100(10): p. 6263–6268.

[66] Yasuda, S., J. Hayashi, and I. Moriya, Genetic constitution for spring growth habit and some other characters in barley cultivars in the Mediterranean coastal regions. Euphytica, 1993. 70(1–2): p. 77–83.

[67] Cockram, J., H. Jones, F.J. Leigh, D. O'Sullivan, W. Powell, D.A. Laurie, and A.J. Greenland, Control of flowering time in temperate cereals: genes, domestication, and sustainable productivity. Journal of Experimental Botany, 2007. 58(6): p. 1231–1244.

[68] Boudiar, R., A. Casas, C. CP, P. Gracia, and E. Igartua, Identification of QTL for agronomic traits contributed by a barley mediterranean landrace. Crop and Pasture

Science, 2016. 67(1): p. 37–46 http://digital.csic.es/bitstream/10261/127755/1/CasasAM_CroPastSci_2016.pdf.

[69] Appels, R., M. Francki, and R. Chibbar, Advances in cereal functional genomics. Functional and Integrative Genomics, 2003. 3(1): p. 1–24.

[70] Faricelli, M.E., M. Valárik, and J. Dubcovsky, Control of flowering time and spike development in cereals: the earliness per se Eps-1 region in wheat, rice, and Brachypodium. Functional and Integrative Genomics, 2010. 10(2): p. 293–306.

[71] Valárik, M., A. Linkiewicz, and J. Dubcovsky, A microcolinearity study at the earliness per se gene Eps-A m 1 region reveals an ancient duplication that preceded the wheat–rice divergence. Theoretical and Applied Genetics, 2006. 112(5): p. 945–957.

[72] Hammer, G.L., G. McLean, S. Chapman, B. Zheng, A. Doherty, M.T. Harrison, E. van Oosterom, and D. Jordan, Crop design for specific adaptation in variable dryland production environments. Crop and Pasture Science, 2014. 65(7): p. 614–626.

[73] Chapman, S.C., G.L. Hammer, and H. Meinke, A sunflower simulation model: I. Model development. Agronomy Journal, 1993. 85(3): p. 725–735.

[74] Imaizumi, T. and S.A. Kay, Photoperiodic control of flowering: not only by coincidence. Trends in Plant Science, 2006. 11(11): p. 550–558.

[75] Dofing, S., Phenological development-yield relationships in spring barley in a subarctic environment. Canadian Journal of Plant Science, 1995. 75(1): p. 93–97.

[76] Aspinall, D. and L. Paleg, Effects of Day-length and Light Intensity on Growth of Barley I. Growth and Development of Apex with a Fluorescent Light Source, 1963, Botanical Gazette Vol. 124, No. 6, p. 429–437.

[77] Laurie, D.A., N. Pratchett, J.H. Bezant, and J.W. Snape, Genetic analysis of a photoperiod response gene on the short arm of chromosome 2 (2H) of Hordeum vulgare (barley). Heredity-LONDON, 1994. 72: p. 619–627.

[78] Börner, A., G. Buck-Sorlin, P. Hayes, S. Malyshev, and V. Korzun, Molecular mapping of major genes and quantitative trait loci determining flowering time in response to photoperiod in barley. Plant Breeding, 2002. 121(2): p. 129–132.

[79] Alqudah, A.M., R. Sharma, R.K. Pasam, A. Graner, B. Kilian, and T. Schnurbusch, Genetic dissection of photoperiod response based on gwas of pre-anthesis phase duration in spring barley. PLoS One, 2014. 9(11): p. e113120.

[80] Laurie, D.A., Comparative genetics of flowering time. Plant Molecular Biology, 1997. 35(1–2): p. 167–177.

[81] Boden, S.A., D. Weiss, J.J. Ross, N.W. Davies, B. Trevaskis, P.M. Chandler, and S.M. Swain, EARLY FLOWERING3 regulates flowering in spring barley by mediating gibberellin production and FLOWERING LOCUS T expression. The Plant Cell, 2014. 26(4): p. 1557–1569.

[82] Casao, M.C., I. Karsai, E. Igartua, M.P. Gracia, O. Veisz, and A.M. Casas, Adaptation of barley to mild winters: a role for PPDH2. BMC Plant Biology, 2011. 11(1): p. 164.

[83] Stracke, S. and A. Börner, Molecular mapping of the photoperiod response gene ea7 in barley. Theoretical and Applied Genetics, 1998. 97(5–6): p. 797–800.

[84] Jarod, A., B. Rollins, M.A. Drosse, S. Mulki, M. Grando, M. Baum, S. Singh, M. Ceccarelli, and v. Korff, Variation at the vernalisation genes Vrn-H1 and Vrn-H2 determines growth and yield stability in barley (Hordeum vulgare) grown under dryland conditions in Syria. Theoretical and Applied Genetics, 2013. 126: p. 2803–2824.

[85] Karsai, I., K. Mészáros, P. Szücs, P. M. Hayes, L. Láng, and Z. Bedö, Effects of loci determining photoperiod sensitivity (Phd-H1) and vernalization response (Sh2) on agronomic traits in the Dicktoo•Morex barley mapping population. Plant Breeding 1999. 118: p. 399—403.

[86] MASWHEAT, Abiotic stress and agronomic traits: vernalization requirement, 2015. http://maswheat.ucdavis.edu/protocols/Vrn/.

[87] von Zitzewitz, J., P. Szucs, J. Dubcovsky, L. Yan, E. Francia, N. Pecchioni, A. Casas, T.H. Chen, P.M. Hayes, and J.S. Skinner, Molecular and structural characterization of barley vernalization genes. Plant Molecular Biology, 2005. 59(3): p. 449–467.

[88] Sasani, S., M.N. Hemming, S.N. Oliver, A. Greenup, R. Tavakkol-Afshari, S. Mahfoozi, K. Poustini, H.-R. Sharifi, E.S. Dennis, and W.J. Peacock, The influence of vernalization and daylength on expression of flowering-time genes in the shoot apex and leaves of barley (Hordeum vulgare). Journal of Experimental Botany, 2009. 60(7): p. 2169–2178.

[89] Hayes, P., F.Q. Chen, A. Corey, A. Pan, T.H. Chen, E. Baird, W. Powell, W. Thomas, R. Waugh, and Z. Bedo, The Dicktoo×Morex Population, in Plant Cold Hardiness. 1997, Springer, New York, US. p. 77–87.

[90] Francia, E., F. Rizza, L. Cattivelli, A. Stanca, G. Galiba, B. Toth, P. Hayes, J. Skinner, and N. Pecchioni, Two loci on chromosome 5H determine low-temperature tolerance in a 'Nure'(winter)×'Tremois'(spring) barley map. Theoretical and Applied Genetics, 2004. 108(4): p. 670–680.

[91] Trevaskis, B., D.J. Bagnall, M.H. Ellis, W.J. Peacock, and E.S. Dennis, MADS box genes control vernalization-induced flowering in cereals. Proceedings of the National Academy of Sciences, 2003. 100(22): p. 13099–13104.

[92] Loukoianov, A., L. Yan, A. Blechl, A. Sanchez, and J. Dubcovsky, Regulation of VRN-1 vernalization genes in normal and transgenic polyploid wheat. Plant Physiology, 2005. 138(4): p. 2364–2373.

[93] Fu, D., P. Szucs, L. Yan, M. Helguera, J.S. Skinner, J. Von Zitzewitz, P.M. Hayes, and J. Dubcovsky, Large deletions within the first intron in VRN-1 are associated with spring growth habit in barley and wheat. Molecular Genetics and Genomics, 2005. 273(1): p. 54–65.

[94] Shitsukawa, N., C. Ikari, S. Shimada, S. Kitagawa, K. Sakamoto, H. Saito, H. Ryuto, N. Fukunishi, T. Abe, and S. Takumi, The einkorn wheat (Triticum monococcum) mutant, maintained vegetative phase, is caused by a deletion in the VRN1 gene. Genes and Genetic Systems, 2007. 82(2): p. 167–170.

[95] Cuesta-Marcos, A., P. Szucs, T.J. Close, T. Filichkin, G.J. Muehlbauer, K.P. Smith, and P.M. Hayes, Genome-wide SNPs and re-sequencing of growth habit and inflorescence genes in barley: implications for association mapping in germplasm arrays varying in size and structure. BMC Genomics, 2010. 11(1): p. 707.

[96] Yan, L., D. Fu, C. Li, A. Blechl, G. Tranquilli, M. Bonafede, A. Sanchez, M. Valarik, S. Yasuda, and J. Dubcovsky, The wheat and barley vernalization gene VRN3 is an orthologue of FT. Proceedings of the National Academy of Sciences, 2006. 103(51): p. 19581–19586.

[97] Dubcovsky, J., D. Lijavetzky, L. Appendino, and G. Tranquilli, Comparative RFLP mapping of Triticum monococcum genes controlling vernalization requirement. Theoretical and Applied Genetics, 1998. 97(5–6): p. 968–975.

[98] Mahfoozi, S., A. Limin, P. Hayes, P. Hucl, and D. Fowler, Influence of photoperiod response on the expression of cold hardiness in wheat and barley. Canadian Journal of Plant Science, 2000. 80(4): p. 721–724.

[99] Yan, L., A. Loukoianov, A. Blechl, G. Tranquilli, W. Ramakrishna, P. SanMiguel, J.L. Bennetzen, V. Echenique, and J. Dubcovsky, The wheat VRN2 gene is a flowering repressor down-regulated by vernalization. Science, 2004. 303(5664): p. 1640–1644.

[100] Slafer, G., Genetic basis of yield as viewed from a crop physiologist's perspective. Annals of Applied Biology, 2003. 142(2): p. 117–128.

[101] Appendino, M. and G.A. Slafer, Earliness per se and its dependence upon temperature in diploid wheat lines differing in the major gene Eps-A m 1 alleles. The Journal of Agricultural Science, 2003. 141(02): p. 149–154.

[102] Kuchel, H., G. Hollamby, P. Langridge, K. Williams, and S. Jefferies, Identification of genetic loci associated with ear-emergence in bread wheat. Theoretical and Applied Genetics, 2006. 113(6): p. 1103–1112.

[103] Zikhali, M., M. Leverington-Waite, L. Fish, J. Simmonds, S. Orford, L.U. Wingen, R. Goram, N. Gosman, A. Bentley, and S. Griffiths, Validation of a 1DL earliness per se (eps) flowering QTL in bread wheat (Triticum aestivum). Molecular Breeding, 2014. 34(3): p. 1023–1033.

[104] Griffiths, S., J. Simmonds, M. Leverington, Y. Wang, L. Fish, L. Sayers, L. Alibert, S. Orford, L. Wingen, and L. Herry, Meta-QTL analysis of the genetic control of ear emergence in elite European winter wheat germplasm. Theoretical and Applied Genetics, 2009. 119(3): p. 383–395.

[105] Faricelli, M., M. Valarik, S. Lewis, L. Appendino, and J. Dubcovsky, Physical map of the Eps-Am1 gene region in Triticum monococcum L. 11th International Wheat Genetics Symposium, 2008. http://ses.library.usyd.edu.au//bitstream/2123/3333/1/P266.pdf.

[106] Slafer, G.A., D.F. Calderini, and D.J. Miralles, Yield Components and Compensation in Wheat: Opportunities for Further Increasing Yield Potential: Increasing Yield. Potential in Wheat, Breaking the Barriers Conference Proceedings of a Workshop Held in Ciudad Obregon, Sonora, Mexico. Edt by M.P. Reynolds, S. Rajaram, and A. McNab., 1996(CIMMYT): p. 101–133.

[107] Hoogendoorn, J., A reciprocal F1 monosomic analysis of the genetic control of time of ear emergence, number of leaves and number of spikelets in wheat (Triticum aestivum L.). Euphytica, 1985. 34(2): p. 545–558.

[108] Slafer, G., J. Araus, C. Royo, and L. Moral, Promising eco-physiological traits for genetic improvement of cereal yields in Mediterranean environments. Annals of Applied Biology, 2005. 146(1): p. 61–70.

[109] Zikhali, M. and S. Griffiths, The effect of Earliness per se (Eps) genes on flowering time in bread wheat. Advances in Wheat Genetics, From Genome to Field. Proceeding of the 12th International Wheat Genetic Symposium, 2015.

[110] Herndl, M., J. White, S. Graeff, and W. Claupein, The impact of Vernalization requirement, photoperiod sensitive and earliness per se on grain protein content of bread wheat. Euphytica, 2008. 163(2): p. 309–320.

[111] Castro, A., P. Hayes, L. Viega, and I. Vales, Transgressive segregation for phenological traits in barley explained by two major QTL alleles with additivity. Plant Breeding, 2008. 127(6): p. 561–568.

[112] Takahashi, R. and S. Yasuda. Genetics of earliness and growth habit in barley. in Proceedings of the International Barley Genetics Symposium. 1971.

[113] Gallagher, L., K. Soliman, and H. Vivar, Interactions among loci conferring photoperiod insensitivity for heading time in spring barley. Crop Science, 1991. 31(2): p. 256–261.

[114] Boyd, W., Li, C.-D., and Grime, C. (2008). Genetic control of heading date in spring barley. The Proceedings of 13th Australian Barley Technical Symposium, Esplanade Hotel, Fremantle, Western Australia, 26–30 August 2007.

[115] APSIM, 2015. http://www.apsim.info/Documentation/Model,CropandSoil/CropModuleDocumentation/AgPasture.aspx.

[116] Keating, B.A., P.S. Carberry, G.L. Hammer, M.E. Probert, M.J. Robertson, D. Holzworth, N.I. Huth, J.N. Hargreaves, H. Meinke, and Z. Hochman, An overview of APSIM, a model designed for farming systems simulation. European Journal of Agronomy, 2003. 18(3): p. 267–288.

[117] CCIA/CSIRO, Projections: climate change in Australia. Technical Report, 2015. 1: p. 1–222.

[118] Anwar, M.R., D. Li Liu, R. Farquharson, I. Macadam, A. Abadi, J. Finlayson, B. Wang, and T. Ramilan, Climate change impacts on phenology and yields of five broadacre crops at four climatologically distinct locations in Australia. Agricultural Systems, 2015. 132: p. 133–144.

[119] Harrison, M., B. Cullen, and R. Rawnsley, Modelling the sensitivity of agricultural systems to climate change and extreme climatic events, Agricultural Systems, 2016, 148: p. 135–148.

[120] Harrison, M.T., F. Tardieu, Z. Dong, C.D. Messina, and G.L. Hammer, Characterizing drought stress and trait influence on maize yield under current and future conditions. Global Change Biology, 2014. 20(3): p. 867–878.

[121] Van Gool, D., L. Vernon, and W. Australia, Potential impacts of climate change on agricultural land use suitability: wheat. 2005: Department of Agriculture.

[122] Phelan, D.C., M.T. Harrison, E.P. Kemmerer, and D. Parsons, Management opportunities for boosting productivity of cool-temperate dairy farms under climate change. Agricultural Systems, 2015. 138: p. 46–54.

[123] Bogard, M., C. Ravel, E. Paux, J. Bordes, F. Balfourier, S. Chapman, J. Le Gouis, and V. Allard, Predictions of heading date in bread wheat (Triticum aestivum L.) using QTL-based parameters of an ecophysiological model. Journal of Experimental Botany, 2014. 65 (20): p. 5849–5865.

[124] Asseng, S., B. Keating, I. Fillery, P. Gregory, J. Bowden, N. Turner, J. Palta, and D. Abrecht, Performance of the APSIM-wheat model in Western Australia. Field Crops Research, 1998. 57(2): p. 163–179.

[125] Jamieson, P., M. Semenov, I. Brooking, and G. Francis, Sirius: a mechanistic model of wheat response to environmental variation. European Journal of Agronomy, 1998. 8(3): p. 161–179.

[126] Yin, X., P.C. Struik, F.A. van Eeuwijk, P. Stam, and J. Tang, QTL analysis and QTL-based prediction of flowering phenology in recombinant inbred lines of barley. Journal of Experimental Botany, 2005. 56(413): p. 967–976.

[127] Meinke, H. and G. Hammer, Climatic risk to peanut production: a simulation study for Northern Australia. Animal Production Science, 1995. 35(6): p. 777–780.

[128] Zheng, B., S. Chapman, J. Christopher, T. Frederiks, and K. Chenu. Predicting heading date and frost impact in wheat across Australia. in 17th Australian Agronomy Conference. 2015.

[129] Zheng, B., S.C. Chapman, J.T. Christopher, T.M. Frederiks, and K. Chenu, Frost trends and their estimated impact on yield in the Australian wheatbelt. Journal of Experimental Botany, 2015. 66 (12): p. 3611–3623.

[130] APSIM, Barley, 2015. https://www.apsim.info/Documentation/Model,CropandSoil/CropModuleDocumentation/Barley.aspx.

[131] Manschadi, A., Z. Hochman, G. Mclean, P. DeVoil, D. Holzworth, and H. Meinke. APSIM-Barley model–Adaptation of a wheat model to simulate barley growth and development. in 13th Australian Agronomy Conference: Perth, Western Australia. 2006.

[132] Eticha, F., H. Grausgruber, and E. Berghoffer, Multivariate analysis of agronomic and quality traits of hull-less spring barley (Hordeum vulgare L.). Journal of Plant Breeding and Crop Science, 2010. 2(5): p. 81–95.

[133] Zeng, X., Genetic variability in agronomic traits of a germplasm collection of hulless barley. Genetics and Molecular Research: GMR, 2015. 14(4): p. 18356.

[134] Nuttall, J. G., G. J. O'Leary, J. F. Panozzo, C. K. Walker, K. M. Barlow and G. J. Fitzgerald. Models of grain quality in wheat-A review. Field Crops Research, 2016. doi:10.1016/j.fcr.2015.12.011

[135] Asseng, S., S. Milroy, and M. Poole, Systems analysis of wheat production on low water-holding soils in a Mediterranean-type environment: I. Yield potential and quality. Field Crops Research, 2008. 105(1): p. 97–106.

[136] Goyne, P., H. Meinke, S. Milroy, G. Hammer, and J. Hare, Development and use of a barley crop simulation model to evaluate production management strategies in north-eastern Australia. Crop and Pasture Science, 1996. 47(7): p. 997–1015.

[137] Asseng, S., F. Ewert, P. Martre, R. Rötter, D. Lobell, D. Cammarano, B. Kimball, M.J. Ottman, G. Wall, and J. White, Rising temperatures reduce global wheat production. Nature Climate Change, 2015. 5(2): p. 143–147.

[138] Barlow, K., B. Christy, G. O'Leary, P. Riffkin, and J. Nuttall. Simulating the impact of extreme heat and frost events on wheat production: The first steps. in 20th International Congress on Modelling and Simulation. Modelling and Simulation Society of Australia and New Zealand, Adelaide, Australia. 2013.

[139] Shorter, R., R. Lawn, and G. Hammer, Improving genotypic adaptation in crops–a role for breeders, physiologists and modellers. Experimental Agriculture, 1991. 27(02): p. 155–175.

[140] Wu, R. and D. O'Malley, Nonlinear genotypic response to macro-and microenvironments. Theoretical and Applied Genetics, 1998. 96(5): p. 669–675.

[141] Bondari, K., Statistical analysis of genotype x environment interaction in agricultural research. Paper SD15, SESUG: The Proceedings of the SouthEast SAS Users Group, St Pete Beach, 2003.

[142] FAO, Plant breeding and farmer participation. (Eds) Ceccarelli, S., Guimaraes, EP and Weltzien, E. 2009.

[143] Garcıa del Moral, L., D.J. Miralles, and G.A. Slafer, Initiation and appearance of vegetative and reproductive structures throughout barley development. Barley Science: Recent Advances from Molecular Biology to Agronomy of Yield and Quality, 2002: p. 243–268.

[144] Ceccarelli, S., W. Erskine, J. Hamblin, and S. Grando, Genotype by environment interaction and international breeding programmes. Experimental Agriculture, 1994. 30(02): p. 177–187.

[145] Löffler, C.M., J. Wei, T. Fast, J. Gogerty, S. Langton, M. Bergman, B. Merrill, and M. Cooper, Classification of maize environments using crop simulation and geographic information systems. Crop Science, 2005. 45(5): p. 1708–1716.

[146] Vargas, M., F.A. van Eeuwijk, J. Crossa, and J.-M. Ribaut, Mapping QTLs and QTL× environment interaction for CIMMYT maize drought stress program using factorial regression and partial least squares methods. Theoretical and Applied Genetics, 2006. 112(6): p. 1009–1023.

[147] Duvick, D.N., The contribution of breeding to yield advances in maize (Zea mays L.). Advances in Agronomy, 2005. 86: p. 83–145.

[148] Hammer, G., Pathways to prosperity: breaking the yield barrier in sorghum. Agricultural Science, 2006. 19(2): p. 16–22.

[149] Bakhsh, A., S. Malik, M. Aslam, U. Iqbal, and A. Haqqani, Response of chickpea genotypes to irrigated and rain-fed conditions. International Journal of Agriculture and Biology (Pakistan), 2007. Volume 9, (4): p. 590–593.

[150] Hunt, J. and N. Poole. Simulating leaf area duration to predict yield response to foliar fungicide in wheat and barley. in Proceedings of 15th Agronomy Conference on Food Security from Sustainable Agriculture. 2010.

[151] Klepper, B., T. Tucker, and B. Dunbar, A numerical index to assess early inflorescence development in wheat. Crop Science, 1983. 23(2): p. 206–208.

[152] SolarTopo, Day length, sunrise and sunset calculator, 2016. http://www.solartopo.com/daylength.htm.

Section 3

Genetic

Chapter 12

Toward a New Paradigm for the Evolution of Developmental and Growth-Pattern Systems in Plants and Animals

Michael M. Lieber

Additional information is available at the end of the chapter

http://dx.doi.org/10.5772/64785

Abstract

Genetically controlled and environmentally responsive mutation as a significant feature of evolution has very likely occurred on different genomic levels. The evolution of developmental and growth-pattern systems in plants and animals could have occurred through a karyotypic mutator system creating controlled, frequent genomic changes on the karyotypic level in response to environmental stresses, such as temperature changes. Such a mutator system generating controlled karyotypic changes at very high frequency in response to stress was discovered. in the fungus, *Aspergillus nidulans*, once classified within the plant kingdom. This mutator system is itself representative of a basic, responsive developmental system producing changes in growth-pattern, morphology, and changes ensuing in a new pattern of differentiation, which are adaptive. Such a developmental, karyotypic mutator system may itself have evolved, through its own self-controlled evolution, into types of complex developmental systems that, through controlled, specific, and minute karyotypic changes during ontogeny, could control patterns of development in plants and animals, integrating different levels of organization. The deeper implications for development and evolution are illustrated, suggesting a new paradigm.

Keywords: model, organism, fungus, environmentally responsive, inner-controlling, karyotypic mutator, mutation, high, frequency, genome, instability, level, stress, accelerated, evolution, evolving, developmental, differentiation, growth, pattern, plants, new, paradigm, consequences

1. Introduction: environmentally responsive mutagenesis on different levels of the genome

Experimental investigations of mutations in various unicellular or simple colonial organisms have revealed the beginnings of a paradigm shift in biology. During the past 27 years, environmentally accommodating or adaptively, reactive, enhanced mutation connected to stressful conditions has been found in bacteria and yeast. In 1969, these were also evident within the unicellular green alga, *Chlamydomonas*, a eukaryote. (See references in [1]. Additional references can be accessed via http://www.googlescholar.com) However, with regard to yeast, there was also a far earlier report [2] of reactively, accommodating mutation to environmental stress. Such enhanced mutations, connected dynamically to stress, allowed for the sudden adaptation or accommodation of the single cells of the organism to changed, stressful situations. In various investigations, such adaptive accommodation via enhanced mutation to a particular nutritional stress allowed single cells in non-growing bacterial colonies to produce adapted, growing clones or sectors, which are referred to as papillae. While in many other investigations pertaining to other types of nutritional stress, such adaptation or accommodation allowed the growth of whole colonies from single cells during the stressful conditions. In 1989, 1990, and 1998, the author showed, through his own work on bacteria under nutritional stress, that the occurrence of a reactively accommodating mutagenesis to stress, which ensues in growing colonies that have accommodated to nutritional stress, is under internal, genetic control or regulation. This demonstrated that such mutation is nonrandom, connected dynamically to the stress, and also demonstrated that the mutagenesis displayed developmental features [3–5]. As pointed out, this indicated the evolution of an inner, mutator capacity, which could have defined the rate of evolution itself.

Earlier in 1967, it was also pointed out by the author [6] that the enhanced occurrence of many types of mutation was nonrandom, being under genetic regulation through internal mutator processes, whose existence in the past could have enhanced the degree of evolution from within. The nonrandomness of enhanced mutations within organisms under stress has become clearly manifested repeatedly in the last 27 years of mutation research. However, during this 27-year period of investigations, the particular, nonrandom mutations studied were only adaptively responsive mutations to nutritional requirements and to the stress of antibiotics. And, in the case of the green alga, the accommodating, reactive, or responding mutagenesis of high degree to stress, occurring within cells on the culture medium, permitted the growth of many colonies of joined cells in the presence of a chemical growth inhibitor within the culture medium. Such a growth inhibitor was another type of chemical stress. These accommodating reactively, enhanced mutations, connected to environmental stress, permitted accommodation or adaptation to stress on the molecular level of genetic organization in unicellular or simple colonial organisms. This would be in contrast to mutations that allow accommodation involving higher levels of organization, such as on the morphogenic level in a multicellular, differentiated organism. This would be a living organism that is comparatively far more complex than bacteria colonies and the colonial algae.

http://dx.doi.org/10.5772/64785

In unique contrast, very frequent mutations at the chromosomal or karyotypic level connected to or in response to physical stress can occur under inner control that lead to adaptive changes in the differentiation of pattern and morphology in olive-green, multicellular, internally organized fungal colonies having a central, crinkled morphology, sparsely populated with conidiophores/conidia, and of reduced growth rate. (See [5] for a brief description and earlier, relevant references.) An adaptively responsive, innerly controlled, greatly enhanced mutation on a higher genomic level determining development was shown to exist many years previously to most of the mutation studies referred to above, with significant implications for evolution, especially for the environmentally responsive evolution of developmental systems.

Investigation with the multicellular, differentiated eukaryotic fungus, *Aspergillus nidulans*, an ascomycete, once considered a lower plant, revealed that very frequent mutations on the karyotypical or chromosomal level of organization were an adaptive response to high-temperature stress. These environmentally responsive, adaptive, karyotypic mutations, a type of controlled instability, resulted in the production of many yellow sectors in each sparsely conidiated, olive-green colony. The sectors were composed of yellow, asexual reproductive structures, the conidiophores, made up of yellow conidia or spores, the means of asexual production. The ensuing production of such mutant sectors manifested itself phenotypically at a higher level as a new type of pattern differentiation through such sectors and morphological change within fungal colonies [7, 8]. This new pattern of differentiation and alteration in morphology relies upon inner-controlled, though environmentally reactive genomic changes. This would be a responding or reactive, inner-controlled hypermutation to stress on the level of the karyotype. These mutagenetic processes were adaptive or accommodating to stress in the following ways: Such inner-determined genomic alterations permitted or developmentally allowed, under temperature stress, large increases in vegetative, yellow spore production within the various, differentiated, mutant yellow sectors. The phenotypic, developmental consequence of or connection to such mutation also greatly increased the growth rate of such, flat, yellow sectors of normal morphology. (Note the photographs at the end of this article.) As will be illustrated, this mutator system is in itself an example or a model of an early developmental system that could have evolved into more complex, developmental systems through its own inner-controlled, adaptively, responsive mutability or instability at the karyotypic level of the genome.

2. Details of the investigation with *Aspergillus nidulans*, a plant-like organism

The fungus investigated, *A. nidulans*, is a normally haploid, eukaryotic ascomycete with eight chromosomes. Its colonies have internally septate hyphae made up of multinucleated cells divided by the respective septa. Without chromosomal rearrangements or new chromosomal configurations within the haploid genome, the fungus produces flat, grass green colonies due to green conidiophores emerging vertically from the multinucleated hyphae composing the complex colonies. The colonies display high growth rates at various temperatures [7, 9]. Colonies with a single, new chromosomal configuration in each of their nuclei have a crinkled morphology and a reduced growth rate, especially at high temperature. The new chromosomal configuration responds self-mutagenically to various temperatures [9]. This fungus produces asexually reproductive spores, which are the source of new fungal colonies. Each spore, a conidium, has a single nucleus.

One particular strain of *A. nidulans* investigated at a high temperature (and at a lesser temperature) has two chromosomal configurations in the haploid genome. The configurations are nonuniform. These configurations have, respectively, partial duplications in trans of chromosomes I (Duplication I) and III (Duplication III). (These resided on respective chromosomal translocations.) *Aspergillus* colonies with these two karyotypic configurations or structures in the haploid genome are much smaller than normal colonies. Unlike normal colonies, those colonies have a morphology which is crinkled. This is especially pronounced at high temperature. These colonies produce far less vegetative spores or conidia at higher temperature, for example, 39.5°C [7, 8]. One of these two karyotypic configurations, defined as Duplication I, has two alleles or genes for conidial color. One allele is for green pigment production, and the other allele is for yellow pigment production. The two color alleles are heterozygous at the same locus within Duplication I. The green allele is dominant to the yellow allele, hence the green or olive-green color of the colonies. That is, these are colonies having green or olive-green conidia and conidiophores within the crinkled area. In some nuclei, a specific region of Duplication I containing the green allele is subject to deletions, resulting in yellow sectors of increased growth rate. The frequency of such deletion from Duplication I, and of corresponding yellow-sector production, is influenced by the other duplication, Duplication III, and by temperature.

Modulated by temperature and the age of the conidia from which colonies are obtained [7], Duplication III controls the degree and pattern of deletion including the green allele on Duplication I. As Duplication III becomes reduced in size as a result of deletions having occurred from it, the reduced Duplication III enhances (at normal culture temperature) the deletion of the genetic region including the green-allele region of Duplication I. A deleted or excised segment, a type of transposition element from Duplication III very likely inserted near the green allele on Duplication I, may trigger, under the control of Duplication III, such deletion. When this occurs under a temperature stress, that is a high temperature for culture growth, this mutagenic, deletional interaction of the two configurations, via a likely transposition process, is enhanced to even a far greater degree in newly regenerated colonies. Moreover, this mutagenic enhancement is clearly regulated, since the improved, yellow sectors, as a consequence of the deletions from Duplication I in many nuclei, all emerge at the same time, as one can see in the photographs. Furthermore, this temporal control of deletion, clearly under the control of the reduced Duplication III, becomes far more pronounced or effective at the stressful, higher temperature [7, 8]. The dampening, epigenetic influence of age-affected conidia on the degree of mutagenic interaction, in cultures obtained from those conidia, is also suppressed epigenetically through this higher temperature [7].

Specifically, at that higher temperature, irrespective of the age state of the conidia producing the fungal colonies, the environmentally accommodating or adaptive results of this extremely high mutagenesis involving controlled deletion, possibly at mitosis, on the

http://dx.doi.org/10.5772/64785

karyotypic level of organization, are adaptively responsive fungal colonies at various levels of organization involving the genomic level. In this regard, each of those colonies respectively and symmetrically produces through the inner-controlled deletions many yellow sectors of increased growth rate. These are sectors also having an abundance of conidia or conidiophores, as well as having a relatively smooth or non-crinkled morphology. These are consequences that are very much accommodating or adaptive to the new temperature situation or induced epigenetic stress. This is especially the situation in the long term within the context of the evolution of new, adaptive differentiation patterns displayed by new adaptive strains of *Aspergillus*. Also, the configurationally, partial duplications, controlling such adaptation, are in effect an adaptively responsive, complex mutator system on the chromosomal or karyotypic level. This is a system that has internal regulation, and one which is environmentally sensitive or reactive. It is a system whose mutagenic behavior is asexually inheritable by means of conidia and sexually transmittable to an F1 generation by means of ascospores [7, 8]. This mutator system has its origin in genomic reorganizations on the karyotypic level. Many types of mutator systems, environmentally sensitive, can be traced to past genetic reorganizations [7, 8].

This situation with *Aspergillus* shows that inner-determined, internally regulated, greatly enhanced karyotypic changes can nevertheless be caused epigenetically by a physical stress, namely high temperature. Such induction occurs in such a manner that controlled karyotypic changes result in adaptive or environmentally accommodating changes on the differentiation/morphological level. This would also be the organizational level of the phenotype. This would be an example of a mutagenic, reactive connection of the karyotypic mutator to physical stress that ensues in a connected, reactively accommodating, differentiation/morphological change to that very physical stress, which permits adaptation. This is a situation that has not been demonstrated before. Such is highly significant as it now shows that morphological and differentiation patterns can be adaptively reactive and connected to environmental stress by means of a stress-induced, influenced mutagenesis involving genomic configurations. These would be developmental, controlling elements on the karyotypic level. Such elements would appear to be regulating the deletion and insertion of smaller transposition elements within the configurations. The physical stress affects, possibly through cytoplasmic and membrane distortions, the inner-controlled mutagenic interaction of the genomic configurations. This occurs in such a way that the regulation becomes enhanced, leading responsively (within one generation) to very frequent, karyotypic-based, controlled alterations in differentiation and morphogenesis in fungal colonies. Such reactive enhancement of mutation on the karyotypic level to environmental stress permits effective, environmentally accommodating alterations on the phenotypic level, which is on the organismal level. The resulting karyotypic alterations have become coextensive with the many yellow mutant sectors within each of a large number of olive-green colonies. And thus, such alterations have become coextensive with a new, adaptive pattern of differentiation and morphogenesis. This represents a reactively or responsively induced, new karyotypic analog of an adaptive differentiation and morphogenesis within a short period, and one connected dynamically to stress.

3. An adaptive phenomenon apparently unique in the history of such investigations

During the many investigations into environmentally accommodating, connected, reactively enhanced mutagenesis, the adaptively responsive phenomenon involving *Aspergillus* was not previously observed. This is especially and specifically the process whereby controlled, very frequent karyotypic change under and through physical stress can be manifested adaptively in a short period as very frequent, adaptive alterations in morphology, growth, and patterns of differentiation within growing, multicellular fungal colonies under stress. It is noted that such responsive adaptation by means of karyotypic mutator systems, whether or not transposition elements are involved, may not be perfect. This is because some karyotypic changes or instabilities could be deleterious. Nevertheless, the types of environmentally responsive mutator systems within *Aspergillus* could have themselves evolved into more effective mutator systems. These would have been systems with developmental features, leading through their evolving, inner-directed changes to more effectively adaptive developmental or morphological solutions to various types of environmental and internally related epigenetic stress.

This phenomenon of environmentally responsive phenotypic change is similar to the phenomenon of the genetic assimilation of induced morphological changes involving environmental stress in *Drosophila*. This was first discovered and investigated by C. H. Waddington in the 1950s [10–12]. In this regard, when developing *Drosophila* embryos are exposed to ether vapor stress treatments or shocks during a specific developmental period, a portion of the *Drosophila* develops two thoraxes with two pairs of wings in adult flies. Within each fly generation exposed to ether stress, developed bithorax flies were inbred or crossed. When after a small number of generations of this inbreeding under stress, a significant proportion of the subsequent progeny resulting from repeated inbreeding for the new morphology, and now free of ether stress, nevertheless still developed the bithorax phenotype as adults. In response to stress, the new morphogenesis has become genetically assimilated or inheritable in a relatively short period. In additional experiments involving inbreeding through a small number of adult generations, other types of morphological changes occurred. These were alterations in wing morphology, eye morphology, and in anal excretory papillae. These were also genetically assimilated following the inbreeding of the adult fly generations, whose embryos responded morphologically to the stress. These were flies whose developing embryos were subject to other types of imposed environmental stresses. These were temperature shocks with regard to wing and eye development and salt treatments of culture media with regard to papilla size.

Not generally investigated was whether or not many of such responsive, genetically assimilated, environmentally responsive morphological changes were adaptive to the environmental stresses in question. However, the genetically assimilated increase in papilla size as a response to salt stress may suggest an inheritable adaptation to the increased salinity in a relatively short period and to any future saline increases. Also, it was not determined whether or not new mutations on the gene and karyotypic level were induced through the imposed environmental

http://dx.doi.org/10.5772/64785

stresses during embryogenesis. Though this situation cannot be ruled out, it should be further investigated. Also, these morphological alterations might have permitted the development of less obvious, internally adaptive and enabling features in complementarity with the evident, genetically assimilated morphological changes. This possibility should also be investigated. In this connection, see [1] regarding what could be enabling mutations in developmental processes.

In later years, additional stress-involved assimilation experiments were performed. This was with a black caterpillar species. Developing embryos of such were subject to heat shocks within each developing, caterpillar generation. As a result, green adults developed during each of a small number of embryo generations subject to heat shock. Subsequently, developing caterpillars eventually became inheritably green without heat shock after a small number of generations through repeated inbreeding of green progeny caterpillars that had developed from exposed embryos in each of those generations [13]. As the authors of this research point out, it is feasible that such inheritably acquired color via heat stress would be adaptive as an effective camouflage in an environment of green, leafy vegetation during the warm season, and thus evolutionally adaptive in a relatively very short period in the context of evolution.

Genetic assimilation of morphological and pattern changes may have played a significant role in evolution. During evolution, such developmental, genetic assimilation of features at the organismal level could have involved some types of environmentally responsive, frequent genomic change on the karyotypic level. They and their effects could have become repeatedly combined through a relatively short period of inbreeding. Hence, this would have accounted for an adaptively responsive assimilation during a relatively, very short period, enabling thereby an accelerated evolution. (In this regard, the adaptive, *Aspergillus* mutator system was generated through inbreeding involving reorganized chromosomes.) With regard to such genetic assimilation of environmentally induced characters, an alteration in genomic organization is indicated [14]. This inheritable or genetic assimilation of environmentally influenced morphological alterations, and less evident, enabled features, could have been the dynamic source for the accelerated, nonlinear evolution of developmental systems in various organisms.

The role of karyotypic mutators in this is quite feasible. This becomes especially feasible in view of the following found with the *Aspergillus* mutator system: one can generate, through an asexual selection from an extremely high mutant-sector, colonial producer at high temperature, a group of colonies with a significantly, further-increased mean frequency of mutant sectors at high temperature compared to the mean mutant-sector frequency of another group of colonies at high temperature [7]. This would certainly suggest a genetic assimilation of a further increased karyotypic mutator effect at high temperature, possibly involving the stabilization of an epigenetic change, itself stressful. And the high-temperature stress would be mutagenic in the context of the inner mutator process, in a way, a nonlinear, epigenetic extension of the mutator process. Occurring in other situations, this could have affected the rate of morphological evolution itself.

4. The evolution of developmental systems due to environmentally responsive genomic changes

Relevantly, a nonlinear rate or burst of karyotypic evolution has been correlated with a high, nonlinear rate of morphological evolution in higher plants and mammals [15]. Karyotypic mutator systems similar to those described in *Aspergillus* may have played a significant role in this [16]. Also, these karyotypic mutator systems may have been mutagenically responding to various environmental and internal stresses. These stresses could have been extremes in temperature and premature aging. The results of these processes may have been corresponding, nearly immediate morphological changes that adaptively accommodated to the new stresses through the environmentally responding mutator system involving the karyotype. This could account for the high, nonlinear rates of morphological evolution of the mammals and of higher plants.

Relevantly, frequent duplications of karyotype leading to polyploidy and corresponding morphological change during plant evolution have been shown to be associated with periods of environmental stress [17]. Polyploidy in plants and general karyotypic change have been very adaptive and have greatly contributed to plant speciation. It cannot be ruled out that such changes in ploidy or karyotype have had, or involved, a developmental, mutator effect, determining in a controlled, specific, and refined manner genomic changes on the karyotypic level. Such mutator systems could have had their origin in those very karyotypic reorganizations. And it is predicted that evidence or indications of this will be discovered in current plants. As long ago as 1940, the geneticist, Richard Goldschmidt, argued that evolution, especially macroevolution, could have involved the responsive or directed generation of mutation on the karyotypic/chromosomal level of organization, ensuing in the sudden occurrence of organisms with new inheritable, developmental, primary patterns [18].

Karyotypic mutator systems may have contributed to and may have themselves become part of the evolution of developmental systems in various organisms. By doing so, they could have determined the very rate or degree of such an evolution [1, 7, 8, 19], consequently enhancing the evolvability of developmental systems. It is likely with regard to the adaptive *Aspergillus* system that those developmental systems would have been the result of an adaptively or environmentally responding, evolving mutator system on the karyotypic level. This would have been a system evolving through its controlled, responding or reactive, connected instability to environmental stress. A consequence of this would have been the evolution in various organisms of even more effectively adaptive, mutator-based developmental systems, wherein inner-controlled, minute karyotypic changes, possibly involving transposons, would have occurred as features of ontogeny. Moreover, such an evolving and integrative mutator system, involving the karyotype, would have determined the very inner evolvability of the evolution of development in various organisms, including and especially in higher plants. In effect, the responsively evolving karyotypic mutator system would be the responsively evolving capacity to evolve developmental systems, the inner-evolving evolvability of evolution. Another avenue for evolution involving mutators might have entailed a modern version of pangenesis proposed in 1967 [6].

5. Likely consequences and possibilities from the evolution of karyotypic mutator systems

Though originating several years ago, experimental studies of *A. nidulans* have nevertheless made explicit, through further examination, a new type of environmentally responding, inner-controlled adaptive mutation of high degree occurring on the karyotypic/chromosomal level. This is an inner-controlled, mutator process, connected to and arising through environmental stress, manifested phenotypically as adaptive changes in growth, differentiation, and morphology. This phenomenon displayed a temporal control reactive and connected to an environmental stress, and it seemed to enable a near-sudden or accelerated adaptive response to a physical stress. Such eventuates when a flexible or plastic accommodation to physical stress becomes necessary. By means of its timing, the phenomenon is adaptively developmental through different levels of organization, from karyotype to pattern differentiation and morphogenesis on the level of the organism. Its inner-regulated temporality is a critical adaptive characteristic of this reactive or responsive phenomenon. This inner capability to eventuate such an adaptively responsive phenomenon and the adaptive, developmental consequences or features is also inheritable.

This adaptively responding process or phenomenon may also be indirectly connected to other environmentally responding alterations in development, which have temporal features and which have become inheritable. Genetic assimilation may be one example of this. There may be other types of environmentally responsive mutator systems that are not evident, yet to be discovered. These may have also played a significant role in the developmental evolution of organisms. Nevertheless, it is likely that many developmental and growth-pattern systems in plants and animals have evolved from a basic, known developmental, karyotypic mutator system, such as the one discovered in *Aspergillus*. Such evolved systems could involve, refined, somatic intrachromosomal recombination. In fact, the process of deletion and transposition in the *Aspergillus* mutator system was proposed to involve specific, somatic intrachromosomal recombination implicating heterochromatin [7, 8].

In various invertebrate animals, controlled, frequent karyotypic changes, such as deletions of heterochromatin, do occur within somatic cells as opposed to germ cells, during development [20–22]. Such deletions or excisions may occur through intrachromosomal recombination [21]. And in certain amphibians, development is known to involve the creation of inheritable, irreversible nuclear (or chromosomal) changes within somatic tissue [see [23]], these changes possibly being deletions. During lymphocyte differentiation in mammals, there is a regulation of genomic rearrangement events in those cells [24]. It is well known that very high-frequency, genomic changes involving somatic hypermutation/intrachromosomal recombination in developmental, immunological tissues (B lymphocytes) occur as a controlled, adaptive response to internal environmental stresses, such as bacteria and viruses or other foreign antigens [25–27]. The developmental consequence is diverse antibody production, which is adaptive. In various plants, there are controlled changes in ploidy in different cells during development [28, 29]. In *Nicotiana*, controlled deletions of heterochromatin in somatic cells, possibly involving intrachromosomal recombination, occur frequently during development, which results in color variegation of the flowers [30]. In maize, some features of development are based on a transposition-insertion-deletion, controlling-element system, with many variations of such [31, 32]. Dr. McClintock, the author of this paradigm-altering research, proposed that many other aspects of maize development could be so based, as well. As in *Aspergillus*, such a system in maize is derived from a chromosomal or karyotypic reorganization. The system in maize and its variations is temperature sensitive. These developmental systems have characteristics suggesting their evolution from responsive karyotypic-based mutators. In view of this, it is likely that other developmental systems occurring through environmentally responsive, changing karyotypes and based on innerly controlled, refined genomic changes will be discovered.

Hence, it is predicted that more and various karyotypic-based mutator systems, responsively generating or leading to frequent adaptive, inheritable changes in differentiation and morphology within short periods, will be detected in various organisms. These mutator systems, through their own inner-controlled instability, may form the basis for the future and rapid evolution of more complex and refined developmental and growth pattern systems, leading to more adaptive and, in many cases, productive organisms. This would include cultivated and nurtured plants used in agriculture and horticulture, but among the harmful, could include organisms that are pathogenic to such plants, as well. The likelihood that the environmentally responsive mutator systems in bacteria, *Aspergillus* and maize, are genetically related through evolution [5] makes this prediction even more likely. The developmental, Ac-Ds controlling-element system in maize [31] is similar to the mutator system in *Aspergillus* [5]. The adaptively responsive phenomenon exhibited by *Aspergillus* (once classified as a lower plant) provides further evidence for the occurrence of various types of adaptively responsive or reactive mutagenesis in many, various organisms. It gives greater likelihood to the conclusions stemming from those previous investigations of adaptively responsive mutagenesis. The developmental, adaptive system in *Aspergillus* makes it more predictable that environmentally responsive, inheritable mutator systems of various types, including those with developmental features, have been a significant parameter in a responsively accelerated, adaptive, developmental evolution of lower organisms, animals, and plants. This would have included the evolution of the progenitors of cultivated crops and of pathogenic organisms.

What appears to be an example of a predicted situation was recently described [33]. When a soil fungus pathogenic to rice was exposed to increasing copper concentrations, which increases are normally toxic to the fungus, and to temperature shocks in other experiments, frequent genomic rearrangements occurred in response to both types of stresses through the agency of transposition elements (TEs). With increasing concentrations of copper in the culture medium, the fungal colonies became resistant to the copper and grew. This was correlated with increased genomic alteration due to the insertion of certain TEs. Furthermore, increased copper resistance was correlated with frequent color changes of the colonies from gray to white. The alterations appeared as white sectors in photographs; morphological alterations were also generated. The fungal colonies adapted to the highest copper concentration displayed dense aerial hyphae. Those colonies were completely white. In earlier investigations by these authors, temperature shocks affected fungal growth and resulted in morphological transitions. These included pigment changes and the production of aerial hyphae. (Personal Communication.)

These responsive, frequent genomic changes to stress occurred over a short period, as indicated by the data. Under field conditions, where there are high concentrations of copper in the soil where the fungus resides, and the soil is very warm due to a tropical environment, this fungus exhibits a high degree of genetic diversity or genetic rearrangements, "suggesting [according to the authors] that high copper content of soil and temperature stress are among the important environmental factors responsible for the high genetic diversity of the pathogen under field conditions." Another implication is that such adaptive, genetic diversity was responsively induced via TEs over a short period.

Though, as noted by the authors, "extensive research over the last several decades has elucidated numerous molecular responses to stress, it is much less known how these translate into organismal–level responses." The authors argue that environmentally responsive TEs indicate such a translation. Might the color and morphological change of the colonies in connection with copper concentration also reflect such a translation? One should recall in this regard that a process very likely involving transposition elements may also have been involved in the adaptively responsive mutator situation in *A. nidulans*. To reiterate, this is a situation where frequent adaptive changes involving color-pattern differentiation, growth, and morphology occurred over a short period. Experimentally supporting such involvement of controlling elements in *Aspergillus*, the transposition of genetic elements that occurred from chromosome to chromosome in *A. nidulans* resulted in morphological and pigment changes within short periods [34]. These transposing elements responsible for those phenotypic changes had their source in a duplication derived from Duplication I. As to whether or not such phenotypic changes, based on such small, mobile, karyotypic segments, were adaptive was, however, not looked into. Yet, experiments conducted by the author showed that high temperature could significantly increase, within a period just over a week, the frequency of generation of this genetically based phenotype [7]. In connection to environmental stresses, karyotypic mutators could have activated and controlled the deletion, transposition, and insertion of genetic elements throughout the genome with developmental effects during the course of their evolution.

The inner-controlled, responsively adaptive processes as described in this article may only be markers or shadows of a deeper, more encompassing adaptive dynamic, whose principle may be independent of scale or level of organization. The elucidation of this process may give a better insight into the translation mentioned above. In this regard, the following questions occur: How and why would the environmentally responsive and innerly controlled karyotypic changes, mediated by TEs, develop into adaptive phenotypes? What are the underlying connections that translate environmental cues or stresses into adaptive, organismal, developmental responses, from phenome to genome and through genome to phenome? The authors, regarding the pathogenic fungus [33], show that the TEs investigated do behave in different ways and are highly specific, responding differently to different environmental clues or stresses. Yet, what occurs through such specificity of action across different levels of organization that ensues in a correct phenotypic adaptation? Further studies of adaptively responding karyotypic mutator systems, which appear to integrate dynamically and specifically those levels, may provide insights into this.

6. Conclusion: toward strengthening the new paradigm with constructive results

Though it appears that karyotypic mutator systems through their own environmentally responsive, inner-controlled instability could have adaptively evolved into many current developmental systems based upon inner-controlled genomic changes, such as those involving transposing genes, it is still not clear in many ways as to how specific adaptive changes on various levels could have been mediated during that evolution. In this regard, could a type of dynamic, epigenetic imprinting due to various stresses, via cellular states, cellular membranes, the cytoskeleton and nuclear matrix, on chromosomal behavior be involved in such specifically responsive adaptations? And could such an imprinting account for a likely responsively accelerated evolution of pathogenic organisms and higher plants through an epigenetic imprinting process regulating and determining lasting karyotypic mutator influences on the very developmentally involved epigenesis itself? Most relevantly, and predictable in this regard, inheritable epigenetic modifications in plants occur due to environmental stresses [35]. Such inheritable, adaptive epigenetic modifications, which the authors refer to as epimutations, are associated with an increased frequency of genomic rearrangements, whose generation appears to be nonrandom.

Such a further evolved, environmentally responsive process could be considered as a transgenerational, environmentally responsive, developmental system, perhaps a variation of genetic assimilation. It would be one manifesting and occurring through dynamic connections across different levels. As far as elucidating the dynamic underlying such specific connections and interconnected adaptations on various levels of organization, including the environmentally responsive, transgenerational epigenetic-karyotypic level, one must look for more interconnected, holistic and imaginative explanations, based on new assumptions. One such assumption could be external forces imprinting stable specificity through instability within and between cellular epigenomes. These explanations and assumptions could and should be elaborated and tested by experiment in order to gain a more complete, empirically based picture and so enable scientists to arrive at a heuristic, universal principle in biology.

Knowing such a principle may enable scientists to counter or reverse the generation and evolution of pathogenic organisms and promote the evolution of pathogenic resistance in crops, as well. Be this as it may, and pointing to aspects of such a principle, environmentally responsive and innerly controlled, adaptively changing karyotypic mutator systems, involving transposons, could have provided the inner dynamic and capacity for various, enhanced macro- and microevolutions of various organisms and their developmental processes over relatively short periods. Using tissue culture methods, the creation and application of such mutator systems in an epigenetic context, involving transmitted energies and stresses, may even become a significant parameter in a near-future evolution. This would occur through the

http://dx.doi.org/10.5772/64785

genetic engineering of more productive, age-resistant plant cultivars with altered, adaptive developmental and growth-pattern systems. These would be developmental changes and features analogous to those generated by the mutator system in *Aspergillus*, a model and primary system. The *Aspergillus* mutator system is an early and significant example (effectively in 1972) of an internally regulated hypermutator system in a relatively complex, multicellular organism enabling quick adaptive responsiveness, on various levels of organization, to new environmentally induced conditions in the organism, and thereby innerly and developmentally evolutionary. The *Aspergillus* mutator system can certainly be seen as being an epigenetic system within a more inclusive one guiding, and cyclically being influenced by inner mutator processes, and one most likely prone to inheritable imprinting.

This would be a type of mutator-based, multilevel, epigenetic system probably forming the evolved basis for many, present-day developmental and growth-pattern systems, at least significant features of such, where controlled genomic change through responsively regulated genetic deletion, transposition, and insertion is involved. Certainly in many cases, gene activation and suppression occur as features of development. Yet, such genetic behavior is dependent on chromosomal configurations or states, such as heterochromatin and methylation. And, predictably, these could very well be epigenetically controlled, as well as controlling, through the environmentally influenced deletion, insertion, and expression of genetic factors, such as transposons, a process representing a type of position effect variegation through regulated intrachromosomal behavior. Modern genetic research has provided supporting evidence of this [e.g., [36]]. It clearly shows that transposition elements play a regulatory role in the development of various organisms, affecting gene expression [37].

As shown in [36], an epigenetic system in a higher plant can induce enhanced, inheritable, and adaptive mutation, through responsive transposon insertion in specific genes, enabling seed germination in response to a chemical stress in culture that inhibits such germination in culture. This is an evolved, mutator-based system controlling development in which transposon activity must also be induced or enabled by the heat treatment of the parental generation prior to and necessary for the chemical-stress induction of the beneficial, adaptive mutations through the stress-responsive insertion of transposons into specific genes of the seed progeny, enabling seed germination. Thus, heat stress itself would be seemingly acting or being utilized within the system in a potentiating-mutagenic, epigenetically adaptive fashion. However, an implicated, controlling methylation of the inserted transposons—where methylation is under the regulation of another genetic region within the system—can inhibit the expression of the adaptive mutations, ensuing in resensitivity to the chemical stress. Subsequent heat treatment reactivates the expression of the beneficial mutations, as well as the expression of adjacent genes, through demethylation of the inserted transposons. The regulated methylation could mask the growth effect of the mutant genes in vivo when conditions would require plant dormancy. Under such cold conditions, as the research implies, the effect of the mutant genes would be nonadaptive but adaptive under warm conditions. The chemical stress is in fact a plant hormone that induces dormancy under cold conditions. In view of this, the evolved, inclusive, and responsive epigenetic control of mutant induction and expression would quickly be able to accommodate plants transgenerationally to changing environmental conditions, allowing for and inhibiting development when, respectively, necessary, and in a heritable fashion. Even though all the adaptive dynamics across different levels of organization have not been clarified, the systems or processes such as these can nevertheless be seen as also contributing to the beginning stages of a new paradigm for mutation and evolution.

A new paradigm encompassing mutation and evolution not only becomes creditable but very feasible. As viewed through this paradigm change, responsively, adaptively, enhanced genetic mutation on various genomic levels can occur, while defining or structuring levels of biological evolution so guided responsively via epigenesis by that mutation. This would be, through mutator processes, an inner-regulated, transgenerational, environmentally responsive, enhanced mutation on different levels to stresses. Thereby, this would have been a mutator-defined, responsive mutation controlling and structuring the rapid and responsively accelerated evolution of organismic, adaptive, developmental capabilities, and their expression. On a deeper, integrated level, the evolution of developmental and growth pattern systems would appear to have an inner, ordering, stabilizing dynamic or component capable of quickly accommodating adaptively to environmental and internally related epigenetic stresses, which tend to destabilize, and which in this context are mutagenic. Thus, evolution itself would appear to be a stabilizing, transgenerational, evolving developmental process, countering destabilization via mutator-controlled, multilevel, responsive mutation, through space-time. This perspective would not only have significant implications for agricultural research, such as crop improvement but could guide medical research, as well.

7. Photographs of *Aspergillus* colonies

The four photographs each show an *Aspergillus* colony having produced through a karyotypic mutator system many mutant yellow sectors in response to a temperature stress. These colonies

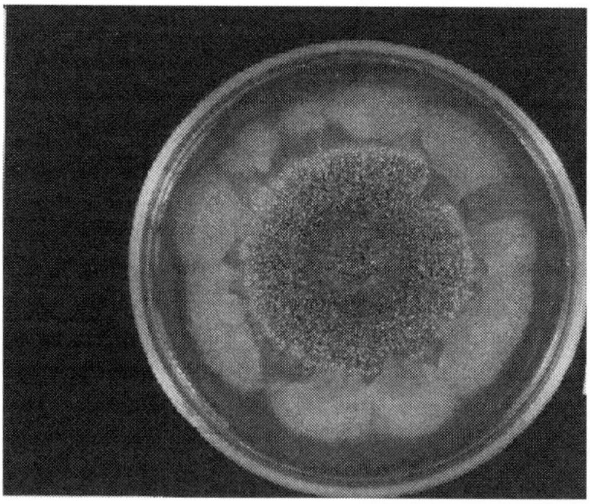

I

http://dx.doi.org/10.5772/64785

came from a large number of colonies displaying the consequences of such mutability producing the adaptive differentiation through the pattern of yellow sectors. Note in photograph IV the two white, mutant sectors displaying a normal morphology and improved growth rate. Their production in this situation, which only occurs in the mutator strain, might have been the result of the insertion of a small genetic element from Duplication III into an epistatic gene on chromosome II that influences pigment production.

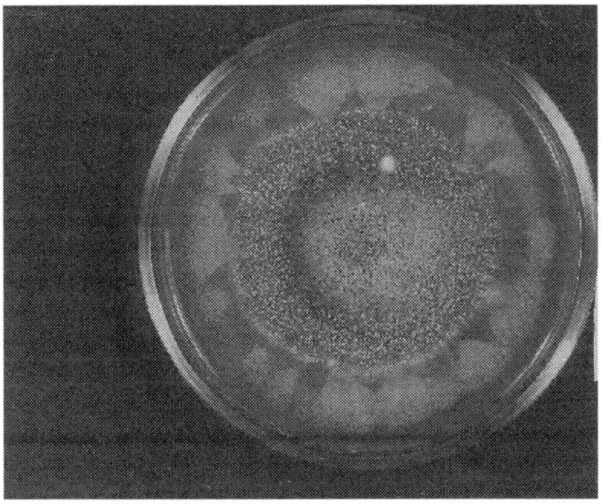

II

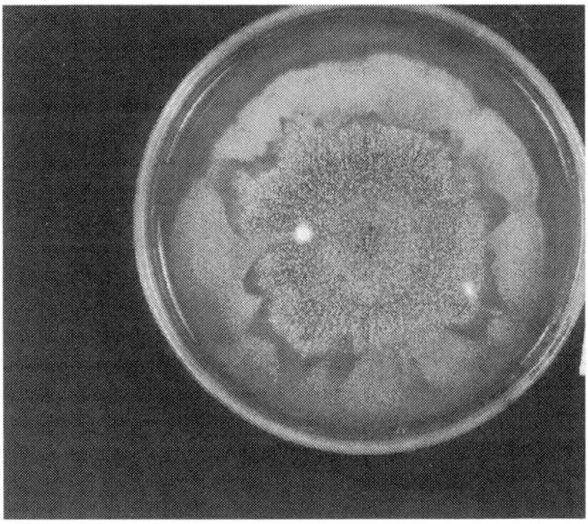

III

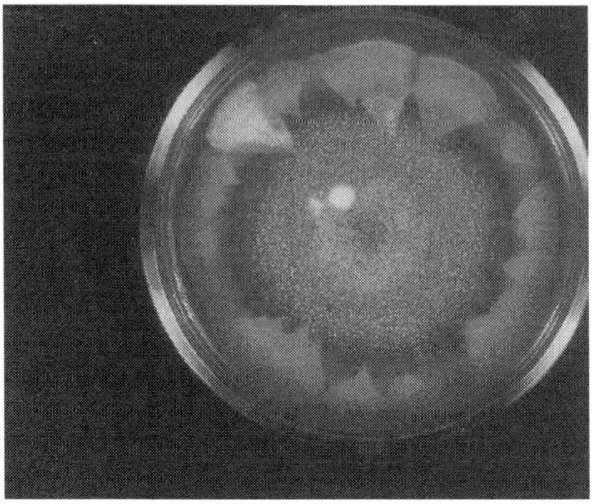

IV

Author details

Michael M. Lieber

Address all correspondence to: michaellieber@juno.com

Genadyne Consulting, Berkeley, CA, USA

References

[1] Lieber, M. (2011). The problem of antibiotic resistant bacteria. The important role of environmentally responsive mutagenesis, its relevance to a new paradigm that may allow a solution. *Theor. Biol. Forum* 104, No. 1: 91–102.

[2] Lindegren, C. (1966). *The Cold War in Biology*, Planarian Press. Ann Harbor, Michigan, USA.

[3] Lieber, M. (1989). New developments on the generation of mutations in *Escherichia coli* lysogens. *Acta Microbiol. Hungarica* 36(4): 377–413.

[4] Lieber, M. (1990). Mutagenesis as viewed from another perspective. *Riv. Biol. B. Forum* 83 (4): 513–522.

[5] Lieber, M. (1998). Environmentally responsive mutator systems: toward a unifying perspective. *Riv. Biol. B. Forum* 91: 425–458.

[6] Lieber, M. (1967). Mutation, development and evolution. Thesis. Institute of Animal Genetics, University of Edinburgh.

[7] Lieber, M. (1972). Environmental and genetic factors affecting instability at mitosis in *Aspergillus nidulans*. Ph.D. Thesis, University of Sheffield.

[8] Lieber, M. (1976). The genetic instability and mutagenic interaction of chromosomal duplications present together in haploid strains of *Aspergillus nidulans. Mutat. Res*. 37: 33–66.

[9] Lieber, M. (1976). The effects of temperature on genetic instability in *Aspergillus nidulans. Mutat. Res*. 34: 94–122.

[10] Waddington, C. H. (1953). Genetic assimilation of an acquired character. *Evolution* 7: 118–126.

[11] Waddington, C. H. (1956). Genetic assimilation of the bithorax phenotype. *Evolution* 10: 1–13.

[12] Waddington, C. H. (1959). Canalization of development and genetic assimilation of acquired characters. *Nature* 183: 1654–1655.

[13] Suzuki, Y. and Nijhout, H.F. (2006). Evolution of a polyphenism by genetic accommodation. *Science* 311: 650–652.

[14] Piaget, J. (1974). *Biology and Knowledge,* University of Chicago Press. Chicago, Illinois, USA.

[15] Wilson, A.C. et al. (1977). Biochemical evolution. *Annu Rev Biochem*. 46: 573–639.

[16] Cherry, L., Lieber, M., and Wilson. A. (1978). Phylogenetic analysis of chromosomal evolution in vertebrates. A Report submitted to the *Proc. Natl. Acad. Sci. USA. Unpublished*.

[17] Vanneste, K., Maere, S., and Van de Peer. Y. (2014). Tangled up in two: a burst of genome duplications at the end of the Cretaceous and the consequences for plant evolution. *Phil. Trans. R. Soc. B*. 369, No. 1648 20130353.

[18] Goldschmidt, R. (1940). *The Material Basis of Evolution,* Yale University Press. New haven, Connecticut, USA.

[19] Lieber, M. (1975). Environmental and genetic factors affecting chromosomal instability at mitosis and the importance of chromosomal instability in the evolution of developmental systems. *Evol. Theory* 1: 97–104.

[20] Goday, C. and Estaban, M.R. (2001) Chromosome elimination in sciarid flies. *Bio Essays* 23 (3): 242–250.

[21] Beerman, S. (1966). A quantitative study of chromatin diminution in embryonic mitoses of *Cyclops furcifer. Genetics* 54: 567–576.

[22] Waddington, C.H. (1956). *Principles of Embryology*, p. 352. G. Allen and Unwin Ltd., London.

[23] Fischberg, M. and Blackler, A.W. (1961). How cells specialize. *Sci. Am.* 205 (3): 124–132.

[24] Alt, F.W. et al. (1986). Regulation of genomic rearrangement events during lymphocyte differentiation. *Immunol.* 89: 5.

[25] Teng, G. and Papavasilion, N. (2007). Immunoglobulin somatic hypermutation. *Ann. Rev. Genet.* 41: 107–120.

[26] Ziqiano, L. et al. (2004). The generation of antibody diversity through somatic hypermutation and class switch recombination. *Genes Dev.* 18: 1–11.

[27] Mange, A. and Mange E. (1990). *Genetics: Human Aspects*, Sinauer Assocs. Sunderland, Mass.

[28] Bino, R.J. et al. (1993). Flow cytometric detection of nuclear replication stages in seed tissue. *Ann. Bot.* 72: 181–187.

[29] Galbraith, D.W. et al. (1991). System endopolyploidy in *Arabidopsis thaliana. Plant Physiol.* 96: 985–989.

[30] Burns, J.A. and Gerstel, D.U. (1967). Flower colour variegation and instability of a block of heterochromatin in *Nicotiana. Genetics* 57, 155.

[31] McClintock, B. (1951). Chromosome organization and genic expression. *Cold Spring Harbor Symp. Quant. Biol.* 16: 13–47.

[32] McClintock, B. (1965). The control of gene action in maize. *Brookhaven Symp. Biol.* 18: 162–203.

[33] Chadha, S. and Sharma, M. (2014). Transposable elements as stress adaptive capacitors induce genomic instability in fungal pathogen *Magnaporthe oryzae*. PLoS One 9, No. 4: 1–14.

[34] Azevedo, J.L. and Roper, J.A. (1970). Mitotic non-conformity in *Aspergillus*: successive and transposable genetic changes. *Genet. Res.* 16: 79–93.

[35] Boyko, A. and Kovalchuk, I. (2011). Genome instability and epigenetic modification—heritable response to environmental stress. *Curr. Opin. Plant Biol.* 14: 260–266.

[36] Ito, H. et al. (2016). A stress-activated transposon in *Arabidopsis* induces transgenerational abscisic acid insensitivity. Published online in *Sci. Rep* 6. Article number 23181.

[37] Pray, L. (2008). Transposons or jumping genes: not junk DNA? *Nat. Educ.* 1 (1): 32.